D1426740

ADVANCE PRAISE FOR

Technology Change and the Rise of New Industries

"In this important book, Jeff Funk examines what it will take to realize the potential of new technologies for innovating out of the economic challenges that we face. He argues that many theories of innovation are incomplete, outdated, or just plain wrong, and that new insights are sorely needed. The basic message is that we have to think of disruption and discontinuity as positive, productive forces that must be embraced."

—Anita M. McGahan, University of Toronto and Author of *How Industries Evolve*

"Jeff Funk is one of the few scholars in the field of technological change who has tried to see the technology system as a whole. This book should be required reading for academics and government decision makers who need to acquaint themselves with this view."

—Robert U. Ayres, INSEAD and International Institute for Applied Systems Analysis

"Without resorting to singular case studies, Funk takes a sophisticated approach to characterizing technological emergence and change, and the role that governments play in developmental trajectories. He builds a descriptive model using a historical approach to reinterpret data from a host of industries. Based on this model, the book takes a daring step to speculate on future technological developments in energy and electronics, providing sobering advice to those who think that government intervention is the panacea for national innovation."

—Phillip Phan, The Johns Hopkins Carey Business School

"In this vitally important advance in the analysis of innovation, Funk explores the limits of learning curves as a mere function of forced or subsidized volumes. Learning has to be real, integrated, and multifaceted in order to benefit from different paths to improvement in cost and performance. Trenchantly demonstrating the need for multidimensional, supply-side innovation in the case of clean energy, he shows the futility of our current demand-side focus."

—George Gilder, Venture Capitalist and Author of the forthcoming *Knowledge and Power: The Information Theory of Capitalism*

"Jeff Funk's provocative elaboration on Giovanni Dosi's notion of technology paradigms calls for a fundamental reexamination of conventional management wisdom about technologies and technology evolutions. In particular, Funk's clear exposition of the supply-side technology dynamics that drive disruptive innovations provides a long overdue corrective to the demand-side story that has been widely advanced by Clayton Christensen, for example. More generally, Funk's framework for analyzing and predicting future technology trajectories establishes a new and essential perspective for both technology strategists and technology policymakers."

—Ron Sanchez, Copenhagen Business School

"Explaining much about innovation that others have ignored, Funk helps us to better understand how improvements in costs and performance occur with new technologies. While the conventional wisdom suggests that costs fall as cumulative production increases, Funk shows us that the reality of this relationship is different and more interesting. For example, technologies that benefit from reductions in scale (e.g., integrated circuits) have seen dramatic advances; finding these kinds of technologies (and products based on them) is a major task for R&D managers."

—Christopher L. Magee, Center for Innovation in Product Development, Massachusetts Institute of Technology

INNOVATION *and* TECHNOLOGY *in the* WORLD ECONOMY

Editor

MARTIN KENNEY
University of California, Davis/Berkeley
Round Table on the International Economy

Other titles in the series:

The Evolution of a New Industry: A Genealogical Approach
ISRAEL DRORI, SHMUEL ELLIS, AND ZUR SHAPIRA

The Science of Science Policy: A Handbook
KAYE HUSBANDS FEALING, JULIA I. LANE, JOHN H. MARBURGER III, AND STEPHANIE S. SHIPP, EDS.

Restoring the Innovative Edge: Driving the Evolution of Science and Technology
JERALD HAGE

The Challenge of Remaining Innovative: Insights from Twentieth-Century American Business
SALLY H. CLARKE, NAOMI R. LAMOREAUX, AND STEVEN W. USSELMAN, EDS.

How Revolutionary Was the Revolution? National Responses and Global Technology in the Digital Era
JOHN ZYSMAN AND ABRAHAM NEWMAN, EDS.

Global Broadband Battles: Why the U.S. and Europe Lag Behind While Asia Leads
MARTIN FRANSMAN, ED.

Ivory Tower and Industrial Innovation: University-Industry Technology Transfer Before and After the Bayh-Dole Act in the United States
DAVID C. MOWERY, RICHARD P. NELSON, BHAVEN N. SAMPAT, AND ARVIDS A. SIEDONIS

Technology Change and the Rise of New Industries

JEFFREY L. FUNK

Stanford Business Books

An Imprint of Stanford University Press
Stanford, California

Stanford University Press
Stanford, California

Special discounts for bulk quantities of Stanford Business Books are available to corporations, professional associations, and other organizations. For details and discount information, contact the special sales department of Stanford University Press. Tel: (650) 736-1782, Fax: (650) 736-1784

Printed in the United States of America on acid-free, archival-quality paper

Library of Congress Cataloging-in-Publication Data

Funk, Jeffrey L., author.
Technology change and the rise of new industries / Jeffrey L. Funk.
pages cm. — (Innovation and technology in the world economy)
Includes bibliographical references and index.
ISBN 978-0-8047-8385-9 (cloth : alk. paper)
1. Technological innovations. 2. Industries. 3. Technology—Economic aspects. I. Title. II. Series: Innovation and technology in the world economy.
HC79.T4F86 2012
338′.064—dc23

2012014298

Typeset by Newgen in 10/12.5 Electra

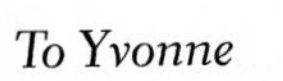

To Yvonne

Contents

Illustrations

Figures

Tables

Acknowledgments

This book benefited from the assistance of many people. Early versions of various chapters benefited from comments by anonymous reviewers on individual papers, feedback on presentations at various conferences and at universities such as Carnegie Mellon and Case Western Reserve, and informal conversations with many people, including Ron Sanchez, Chris Tucci, Phil Phan, Brian Arthur, and Dick Lipsey. For the actual book manuscript, Nuno Gil, Jeroen van der Bergh, and Chris Magee provided written and verbal feedback. Special thanks to Chris for closely reading several chapters and identifying many key issues. Three anonymous reviewers provided many insightful comments, and some of these led to very significant improvements to the manuscript. I would also like to thank Margo Fleming and Jessica Walsh, at Stanford University Press, and Martin Kenney, as the series editor, for their help in turning the manuscript into a final book. Most of all, I would like to thank my wife Yvonne for her constant support.

Technology Change and the Rise of New Industries

1

Introduction

The U.S. and other governments spend far more money subsidizing the production of clean energy technologies, such as electric vehicles, wind turbines, and solar cells, than they do on clean energy research and development (R&D).[1] Why? A major reason is that many believe that costs fall as a function of cumulative production in a so-called learning or experience curve, and thus stimulating demand is the best way to reduce costs. According to such a curve, product costs drop a certain percentage each time cumulative production doubles as automated manufacturing equipment is introduced and organized into flow lines.[2] Although such a learning curve does not explicitly exclude activities performed outside of a factory, the fact that learning curves link cost reductions with cumulative production focuses our attention on the production of a final product and implies that learning gained outside of a factory is either unimportant or is driven by that production.

But is this true? Are cumulative production and its associated activities in a factory the most important sources of cost reductions for clean energy or any other technology for that matter? Among other things, this book shows that most improvements in wind turbines, solar cells, and electric vehicles are being implemented outside of factories and that many of them are only indirectly related to production. Engineers and scientists are increasing the physical scale of wind turbines, increasing the efficiencies as well as reducing the material thicknesses of solar cells,[3] and improving the energy storage densities of batteries for electric vehicles, primarily in laboratories and not in factories. This suggests that increases in production volumes, particularly those of existing technologies, are less important than increases in spending on R&D (i.e., supply-side approaches)—an argument that Bill Gates[4] and other business leaders regularly make. Although demand and thus demand-based subsidies do encourage R&D,[5] only a small portion of these subsidies will end up funding R&D activities.

Should this surprise us? Consider computers (and other electronic products such as mobile phones[6]). The implementation of automated equipment and its organization into flow lines in response to increases in production volumes are not the main reasons for the dramatic reduction in the cost of computers over the last 50 years. The cost of computers dropped primarily for the same reason that their performance rose: continuous improvements in integrated circuits (ICs). Furthermore, improvements in the cost and performance of ICs were only partly from the introduction of automated equipment and its organization into flow lines. A much more important cause was large reductions in the *scale* of transistors, memory cells, and other dimensional features, where these reductions required improvements in semiconductor-manufacturing equipment. This equipment was largely developed in laboratories, and these developments depended on advances in science; their rate of implementation depended more on calendar time (think of Moore's Law) than on cumulative production volumes of ICs.[7]

NEW QUESTIONS AND NEW APPROACHES

We need a better understanding of how improvements in cost and performance emerge and of why they emerge more for some technologies than for others, issues that are largely ignored by books on management (and economics). While most such books are about innovative managers and organizations, and their flexibility and open-mindedness, they don't help us understand why some technologies experience more improvements in cost and performance than do others. In fact, they dangerously imply that the potential for innovation is everywhere and thus all technologies have about the same potential for improvement.

Nothing can be further from the truth. ICs, magnetic disks, magnetic tape, optical discs, and fiber optics experienced what Ray Kurzweil calls "exponential improvements" in cost and performance in the second half of the 20th century, while mechanical components and products assembled from them did not.[8] Mobile phones, set-top boxes, digital televisions, the Internet, automated algorithmic trading (in hedge funds, for example), and online education also experienced large improvements over the last 20 years because they benefited from improvements in the previously mentioned technologies. A different set of technologies (e.g., steam engines, steel, locomotives, and automobiles) experienced large improvements in both cost and performance in the 18th and 19th centuries. An understanding of why some technologies have more potential for improvements than do others is necessary for firms, governments, and organizations to make good decisions about clean energy and new technologies in general.

We also need a better understanding of how science and the characteristics of a technology determine the potential of new technologies. Although there is a large body of literature on how advances in science facilitate advances in technology in the so-called linear model of innovation,[9] many of these nuances are ignored once learning curves and cumulative production are considered. For example, improvements in solar cell efficiency and reductions in material thicknesses involve different sets of activities, and the potential for these improvements depends on the types of solar cells and on levels of scientific understanding for each type. Lumping together the cumulative production from different types of solar cells causes these critical nuances to be ignored and thus prevents us from implementing the best policies.

Part of the problem is that we don't understand what causes a time lag (often a long one) between advances in science, improvements in technology that are based on these advances, and the commercialization of technology. And without such an understanding, how can firms and governments make good decisions about clean energy? More fundamentally, how can they understand the potential for Schumpeter's so-called creative destruction and new industry formation? A new industry is defined as a set of products or services based on a new concept and/or architecture where these products or services are supplied by a new collection of firms and where their sales are significant (e.g., greater than $5 billion). According to Schumpeter, waves of new technologies (which are often based on new science) have created new industries, along with opportunities and wealth for new firms, as they have destroyed existing technologies and their incumbent suppliers.

This is a book about why specific industries emerge at certain moments in time and how improvements in technologies largely determine this timing. For example, why did the mainframe computer industry emerge in the 1950s, the personal computer (PC) industry in the 1970s, the mobile phone and automated algorithmic trading industries in the 1980s, the World Wide Web in the 1990s, and online universities in the 2000s? On the other hand, why haven't the personal flight, underwater, and space transportation industries emerged, in spite of large expectations for them in the 1960s?[10] Similarly, why haven't large electric vehicle, wind, and solar industries yet emerged, or when will such industries emerge that can exist without subsidies?

Parts of these questions concern policies and strategies. When did governments introduce the right polices and when did firms introduce the right strategies? But parts also involve science and technology, and, as mentioned previously, they have been largely ignored by management books on technology and innovation,[11] even as the rates of scientific and technological change have accelerated and the barriers to change have fallen.[12] When was our understanding of scientific phenomena or the levels of performance and cost for the relevant technologies sufficient for industry formation to occur? We need better answers to these kinds of questions in order to complement research on government policies and firms' R&D strategies. For example, understanding the factors that impact on the timing of scientific, technical, and economic feasibility can help firms create better product and technology road maps, business models, and product introduction strategies. They can help entrepreneurs understand when they should quit existing firms and start new ones.[13] And they can help universities better teach students how to look for new business opportunities and address global problems; such problems include global warming, other environmental emissions, the world's dependency on oil and minerals from unstable regions, and the lack of clean water and affordable housing in many countries.

Some of the problems that arise when firms misjudge the timing of economic feasibility can be found in the mobile phone industry. In the early 1980s, studies concluded that mobile phones would never be widely used, while in the late 1990s studies concluded that the mobile Internet was right around the corner. Some would argue that we underestimated the importance of mobile communication, but I would argue that these studies misjudged the rate at which improvements in performance and cost would occur. The 1980s studies should have been asking what consumers would do when Moore's Law made handsets free and talk times less than 10 cents a minute. The 1990s studies should have been addressing the levels of performance and cost needed in displays, microprocessor and memory ICs, and networks before various types of mobile Internet content and applications could become technically and economically feasible.[14]

Chapters 2 and 3 (Part I) address the potential of new technologies using the concept of *technology paradigm* primarily advanced by Giovanni Dosi.[15] Few scholars or practitioners have attempted to use the technology paradigm to assess the potential of new technologies or to compare different ones.[16] One key aspect of this paradigm is geometrical scaling, which is a little-known idea initially noticed in the chemical industries (and in living organisms).[17] Part I shows how a technology paradigm can help us better understand the potential for new technologies where technologies with a potential for large improvements in cost and performance often lead to the rise of new industries. Part I and the rest of this book also show how implementing a technology and exploiting the full potential of its technology paradigm require advances in science and improvements in components.

One reason for using the term "component" is to distinguish between components and systems in what can be called a "nested hierarchy of subsystems."[18] Systems are composed of subsystems, subsystems are composed of components, and components may be composed of various inputs including equipment and raw materials. This book will just use the terms *systems* and *components* to simplify the discussion. For example, a system for producing integrated circuits is composed of components such as raw materials and semiconductor-manufacturing equipment.

TECHNOLOGICAL DISCONTINUITIES AND A TECHNOLOGY PARADIGM

A technology paradigm can be defined at any level in a nested hierarchy of subsystems, where we are primarily interested in large changes in technologies, or what many call technological discontinuities. These are products based on a different set of concepts and/or architectures from that of existing products, and they are often defined as the start of new industries.[19] For example, the first mainframe computers, magnetic tape–based playback equipment, and transistors (like new services such as automated algorithmic trading and online universities) were based on a different set of concepts than were their predecessors: punch card equipment, phonograph records, and vacuum tubes, respectively. On the other hand, minicomputers, PCs, and various forms of portable computers only involved changes in architectures.

Building from Giovanni Dosi's characterization and using an analysis of many technologies (See the Appendix for the research methodology), Chapter 2 and the rest of this book define a technology paradigm in terms of (1) a technology's basic concepts or principles and the trade-offs that are defined by them; (2) the directions of advance within these trade-offs, where advance is defined by a technological trajectory (or more than one);[20] (3) the potential limits to trajectories and their paradigms; and (4) the roles of components and scientific knowledge in these limits.[21] Partly because this book is concerned with understanding when a new technology might offer a superior value proposition, Chapter 2 focuses on the second and third items and shows how there are four broad methods of achieving advances in performance and cost along technological trajectories: (1) improving the efficiency by which basic concepts and their underlying physical phenomena are exploited; (2) radical new processes; (3) geometrical scaling; and (4) improvements in "key" components.

In doing so, Chapter 2 shows how improvements in performance and/or price occur in a rather smooth and incremental manner over multiple generations of discontinuities. While some argue that these improvements can be represented by a series of S-curves where each discontinuity initially leads to *dramatic* improvements in performance-to-price ratios,[22] this and succeeding chapters show that such dramatic changes in the rates of improvement are relatively rare. Instead, this book's analyses suggest that there are smooth rates of improvement that can be characterized as incremental over multiple generations of technologies, and that these incremental improvements in a technological trajectory enable one to roughly understand near-term trends in performance and/or price/cost for new technologies.

GEOMETRICAL SCALING

Chapter 3 focuses on geometrical scaling as a method of achieving improvements in the performance and cost of a technology. Geometrical scaling refers to the relationship between the geometry of a technology, its scale, and the physical laws that govern it. As others describe it, the "scale effects are permanently embedded in the geometry and the physical nature of the world in which we live."[23]

As a result of geometrical scaling, some technologies benefit from either large increases (e.g., engines or wind turbines) or large reductions (e.g., ICs) in physical scale. For example, consider the pipes and reaction vessels that make up chemical plants, which benefit from increases in scale While economies of scale generally refer to amortizing a fixed cost over a large volume, at least until the capacity of a plant is reached, geometrical scaling refers to the fact that the output from pipes varies as a function of one dimension (radius) squared whereas the costs of pipes vary as a function of this dimension (radius) to the first power. Similarly, the output from a reaction vessel varies as a function of one dimension (radius) cubed whereas the costs of the reaction vessels vary as a function of one dimension (radius) squared. This is why empirical analyses have found that the costs of chemical plants rise only about two-thirds for each doubling of output and thus increases in the scale of chemical plants have led to dramatic reductions in the cost of many chemicals.[24]

Other technologies benefit from *reductions* in scale. The most well-known examples of this type of geometrical scaling can be found in ICs, magnetic disks and tape, and optical discs, where reducing the scale of transistors and storage regions has led to enormous improvements in the cost and performance of these technologies.[25] This is because reductions in scale lead to improvements in both performance and costs. For example, placing more transistors or magnetic or optical storage regions in a certain area increases speed and functionality and reduces both the power consumption and size of the final product, which are typically considered improvements in performance for most electronic products; they also lead to lower material, equipment, and transportation costs. The combination of increased performance and reduced costs as size is reduced has led to exponential improvements in the performance-to-cost ratio of many electronic components.

Like Chapter 2, Chapter 3 and other chapters show how geometrical scaling is related to a nested hierarchy of subsystems. Chapter 3 demonstrates that benefiting from geometrical scaling in a higher-level "system" depends on improvements in lower-level supporting "components,"[26] and that large benefits from geometrical scaling in a lower-level "key component" can drive long-term improvements in the performance and cost of a higher-level "system." In the second instance, these long-term improvements may lead to the emergence of technological discontinuities in systems, particularly when the systems do not benefit from increases in scale. Part II shows how exponential improvements in ICs and magnetic storage densities led to discontinuities in computers and magnetic recording and playback equipment, as well as in semiconductors. Chapter 9 explores this for other systems.

In fact, most of the disruptive innovations covered by Clayton Christensen, who many consider to be the guru of innovation,[27] benefit from geometrical scaling (and experience exponential improvements) in either the "system" or a key "component" in the system. This suggests that there is a "supply-side" aspect to Christensen's theory of disruptive innovation that is very different from his focus on the demand side of technological change. While his theory suggests to some that large improvements in performance and costs along a technological trajectory *automatically* emerge once a product finds a low-end niche, and so finding the low-end niche is the central challenge of creating disruptive innovations,[28] Chapters 3 and 4 show how geometrical scaling explains why some low-end technological discontinuities became disruptive innovations and why these low-end technological discontinuities initially emerged. A search for potentially disruptive technologies, then, should consider the extent to which a system or a key component in it can benefit from rapid rates of improvement through, for example, geometrical scaling.

Some readers may find the emphasis on supply-side factors in Chapters 2 and 3 (Part I) to be excessive and thus may classify the author as a believer in so-called technological determinism. Nothing could be further from the truth. I recognize that there is an interaction between market needs and product designs, that increases in demand encourage investment in R&D, and that the technologies covered in this book were "socially constructed."[29] The relevance of this social construction is partly reflected in the role of new users in many of the technological discontinuities covered in Part II, where these new users and changes in user needs can lead to the rise of new industries.[30] For example, the emergence of industries represented by microbreweries and artisanal cheeses is more the result of changes in consumer taste than of changes in technology. Some of these changes come from rising incomes that have led to the emergence of many industries serving the rich or even the super rich. When the upper 1 percent of Americans receives 25 percent of total income, many industries that cater to specialized consumer tastes will naturally appear.[31]

This book focuses on supply-side factors because industries that have the potential to significantly enhance most lives or improve overall productivity require dramatic improvements in performance and cost. As Paul Nightingale says about Giovanni Dosi's theory of technology paradigms, drawing on the research of Nathan Rosenberg and David Mowery,[32] "'Market pull' theories are misleading, not because they assume innovation processes respond to market forces but because they assume that the response is *unmediated*. As a consequence, they cannot explain why so many innovations are not forthcoming despite huge demand, nor why innovations occur at particular moments in time and in particular forms."[33] For example, the world *needs* inexpensive solar, wind, and other sources of clean energy, and large subsidies are increasing demand and R&D spending for them. But even with these large subsidies, significant improvements in cost and performance will not be forthcoming if such technologies do not have the potential for dramatic reductions in cost. And if they don't have such a potential, the world needs to look for other solutions.

A second reason for focusing on supply-side factors is that unless we understand the technological trajectories and the factors that have a direct impact on them, such as scaling, how can we accelerate the rates of improvement in cost and performance? Since much of the management literature on learning primarily focuses on the organizational processes involved with learning, it implies that organizational issues have a bigger impact on the potential for improving costs and performance than does the nature of the technology.[34] Thus, while this literature implies that solving energy and environmental problems is primarily an organizational issue, geometrical scaling and the other three methods of achieving advances in performance and cost remind us that the *potential* for improving cost and performance depends on a technology's characteristics.[35] Without a *potential* for improvements, it would be difficult for organizational learning to have a large impact on the costs and performance of a technology, no matter how innovative an organization is.

THE TIMING OF TECHNOLOGICAL DISCONTINUITIES

Chapters 4 through 6 (Part II) analyze technological discontinuities in part because these discontinuities often form the basis for new industries. For example, the first mainframe computers, minicomputers, personal computers, personal digital assistants, audiocassette players, videocassette recorders, camcorders, memory ICs, and microprocessors, as well as automated algorithmic trading and online education, are typically defined in this way. Like other discontinuities, they were based on a different set of concepts and/or architectures than were existing products. The characterization of a system's architecture is also considered important because the ability to characterize a system's concept partly depends on one's ability to characterize its potential architectures.

But what determines the timing of these discontinuities? Since the characterization of a concept or architecture and an understanding of the relevant scientific phenomenon usually precede the commercialization of a technology, we can look at the timing of technological discontinuities in relation to them. How long before the emergence of technological discontinuities were the necessary concepts and/or architectures characterized? And why is there a time lag and in many cases a long one, between a characterization of these concepts and architectures and both the commercialization and diffusion of the technology?[36]

These questions are largely ignored by academic researchers. While there is wide agreement on the descriptions and timing of specific technological discontinuities, most research focuses on the existence and reasons for incumbent failure and in doing so mostly treats these discontinuities as "bolts of lightning." For example, the product life cycle cyclical, and disruptive models of technological change do not address the sources of technological discontinuities; instead, their emphasis on incumbent failure implies that any time lag is due to management (e.g., cognitive) failure.[37]

But do we really believe that management failure for either cognitive or organizational reasons is why it took more than 100 years to implement Charles Babbage's computing machine in spite of early government funding?[38] Although Babbage defined the basic concept of the computer in the 1820s and subsequently built a prototype, general-purpose computers did not emerge until the 1940s or diffuse widely in developed countries until the 1980s. Is this time lag merely due to narrow-minded managers and policy makers, or is something else going on? More important, in combination with a theory that technological discontinuities initially experience dramatic improvements in performance and price, an emphasis on incumbent failure as the main reason for a long time lag suggests that there are many technological discontinuities with a potential for *dramatic* improvements in performance and costs just waiting to be found. According to this logic, if only managers and policy makers could overcome their cognitive limitations, firms and governments could find new technologies that would *quickly* replace existing ones and thus solve big problems such as global warming.

This book disagrees with such an assessment and shows how the timing of many discontinuities can be analyzed. Building from research done by Nathan Rosenberg and his colleagues on the role of complementary technologies in the implementation of new ones,[39] Part II shows that insufficient components were the reason for the time lag between the identification and characterization of concepts and architectures that formed the basis of technological discontinuities and the commercialization (and diffusion) of these discontinuities.[40] Chapters 4, 5, and 6 present a detailed analysis of the discontinuities in computers, magnetic recording and playback equipment, and semiconductors, respectively. One reason for choosing these "systems" is that few argue that there were market failures for discontinuities in them, unlike the discontinuities of more "complex network" systems such as broadcasting or mobile phones, which are addressed in Part III.[41] A second reason is that there have been many discontinuities in these and related systems, and thus there are many "data points" to analyze.[42] Finally, the time lag for each discontinuity in these systems was primarily due to one or two insufficient components, which is very different from the mechanical sector, where novel combinations of components have probably played a more important role than have improvements in one or two components individually.[43] Partly because it is possible to design many of these systems in a modular way,[44] their performance was primarily driven by improvements in "key" components (which is the fourth broad method of achieving advances in a system's performance and costs), and these improvements also drove the emergence of system discontinuities.

For example, the implementation of minicomputers, personal computers, and most forms of portable computers primarily depended on improvements in one type of component, the IC, as the discontinuities were all based on concepts and architectures that had been characterized by the late 1940s.[45] Similarly, the implementation of various discontinuities in magnetic-based audio and video recording equipment primarily depended on improvements in one type of component, the magnetic recording density of tape, as these discontinuities were all based on concepts and architectures that had been characterized by the late 1950s. In other words, in spite of the increasing variety of components that can be combined in many different ways, improvements in a single type of component had a larger impact on the emergence of these discontinuities (and on the performance of these systems) than did so-called novel combinations of multiple components (or technologies). This conclusion enables us to go beyond the role of complementary technologies in the time lag to analyze the specific levels of performance that were needed in single types of components before new systems—that is, discontinuities—could be implemented.

SYSTEMS, COMPONENTS, AND DISCONTINUITIES

Chapters 4, 5, and 6 analyze the impact of improvements in a single type of component on the emergence of discontinuities in systems in two different ways, where both of these ways are facilitated by the smooth and incremental manner in which improvements in performance and/or price have been occurring.

First, building from the role of trade-offs in technology paradigms and marketing theory,[46] these chapters show how improvements in components have changed the trade-offs that suppliers (i.e., designers) and users make when they consider systems and how this change leads to the emergence of discontinuities. Technology paradigms define a set of trade-offs between price and various dimensions of performance, which suppliers consider when they design or compare systems, while users make trade-offs between price and various dimensions of performance. In both cases, improvements in components can change the way these trade-offs are made by both suppliers and users.

Second, economists use the term "minimum threshold of performance" to refer to the performance that is necessary before users will consider purchasing a system.[47] For example, users would not purchase a PC until PCs could perform a certain number of instructions per second in order to process specific software applications. When a single type of component such as a microprocessor has a large impact on the performance of a system such as a PC, a similar threshold exists for the components in these systems. Thus, PCs could not perform a certain number of instructions per second until a microprocessor could meet certain levels of performance.

Part II draws a number of conclusions from these analyses. First, the new concepts or architectures that form the basis of discontinuities in systems were known long before the discontinuities were implemented. In other words, the characterization of concepts or architectures was usually not the bottleneck either for the discontinuities or for the creation of the industries that many of these discontinuities represent. Instead, the bottleneck was one or two types of components that were needed to implement the discontinuities. Thus, improvements in components can gradually make new types of systems (i.e., discontinuities) possible, and the thresholds of performance (and price) that are needed in specific components before a new system is economically feasible can be analyzed.

Second, finding new customers and applications, which partly reflect heterogeneity in customer needs,[48] can reduce the minimum thresholds of performance for the components needed to implement discontinuities. Chapters 4, 5, and 6 provide many examples of how new customers and applications (and methods of value capture) enabled discontinuities to be successfully introduced before the discontinuities provided the levels of performance and/or price that the previous technology provided. In other words, these new customers, applications, and method of value capture reduced the minimum thresholds of performance for these systems and their key components. However, although this was important from the standpoint of competition between firms, the impacts on these thresholds of new customers, applications, and methods of value capture (and the heterogeneity in customer needs they reflected) were fairly small when compared to the many orders of magnitude in system performance that came from improvements in component performance.

Third, one reason that discontinuities emerged in computers and in magnetic recording and playback equipment is that they did not benefit from geometrical scaling to the extent that their components did. ICs and magnetic recording density experienced exponential improvements in cost and performance because they benefited from dramatic reductions in scale (i.e., geometrical scaling). However, since computers and magnetic recording systems do not benefit from increases in scale (as do, for example, engines), it was natural that smaller versions emerged and replaced larger ones.

Fourth, the demand for many of these improvements in components was initially driven by other systems and/or industriesThis enabled the new systems to receive a "free ride" from existing industries as improvements in components "spilled over" and made the new industries possible. This provides additional evidence that the notion of cumulative production driving cost reductions is misleading and impractical, a point that others such as William Nordhaus have made using different forms of analysis.[49] Not only is it by definition impossible for learning curves to help us understand when a potential discontinuity (one not yet produced) might provide a superior value proposition; Part II shows how improvements in components (e.g., ICs) gradually made new discontinuities economically feasible where the demand for these components was coming from other industries.

CHALLENGES FOR FIRMS AND GOVERNMENTS

Chapters 7 and 8 of Part III address a different set of questions: those that concern the challenges for firms and governments with respect to new industries.

While Parts I and II focus on when a discontinuity might become economically feasible, and thus imply that firms easily introduce and users easily adopt new technologies, Chapters 7 and 8 summarize the complexities of new industry formation and thus the challenges for firms and governments. These complexities may cause the diffusion of new technologies to be delayed, or they may enable new entrants or even new countries to dominate an industry whose previous version was dominated by other countries.

Chapter 7 focuses on competition between firms. Incumbents often fail when technological discontinuities emerge and diffuse, particularly when these discontinuities destroy an incumbent's capabilities.[50] New technologies can destroy a firm's capabilities in many areas, including R&D, manufacturing, marketing, and sales, where this destruction may be associated with the emergence of new customers. For example, Christensen argues that incumbents often fail when a low-end innovation displaces the dominant technology (thus becoming a disruptive innovation) largely because it initially involves new customers and serving them requires new capabilities.[51] Helping firms analyze the timing of technological discontinuities, which is the subject of Part II, can help them identify and prepare for these discontinuities through, for example, identifying the appropriate customers and creating the relevant new capabilities to serve them.

Other research has found that the total number of firms declines quickly following the emergence of a technological discontinuity in some industries more than in other industries where the number of firms is a surrogate for the number of opportunities.[52] This decline occurs through mergers, acquisitions, and exits, in what many call a "shakeout." The occurrence of a shakeout depends on whether large firms have advantages over smaller firms through economies of scale in operations, sales, and/or R&D. For example, economies of scale in R&D (or other activities) favor firms with large sales because they can spend more on total R&D than can firms with fewer sales. Initially, greater spending on R&D leads to more products and more products lead to more sales; this positive feedback leads to larger firms dominating an industry, where smaller firms are acquired or exit the industry.[53]

Chapter 7 focuses on these issues in more detail and on how two factors, the number of submarkets and the emergence of vertical disintegration, affect the importance of economies of scale and thus the number of opportunities for new entrants. The existence of submarkets can reduce the extent of economies of scale in R&D when each submarket requires different types of R&D; thus the existence of submarkets can prevent a shakeout. This enables a larger number of firms, including entrepreneurial start-ups, to exist in an industry or sector with many submarkets than in one with few submarkets.

Vertical integration permits the late entry of firms, sometimes long after a shakeout has occurred. Furthermore, since vertical disintegration can lead to a new division of labor in an economy in which a set of new firms provides new types of products and services, it, too, can lead to the rise of new industries. While Chapter 7 primarily focuses on the emergence of high-technology industries such as computer software, peripherals, and services, and semiconductor foundries and design houses, vertical disintegration has also led to the formation of less high-tech, albeit large, industries such as janitorial services, credit collection, and training services.[54]

Chapter 8 focuses on how the challenges for firms and governments vary by type of industry, using a typology of industry formation. Whereas industries might emerge from either vertical disintegration or technological discontinuities, most of the examples in Chapter 8 are for those that emerged from the latter. The typology focuses on system complexity and whether a critical mass of users or complementary products is needed for growth to occur. Although the formation of most new industries depends on when a new technology becomes economically feasible and thus provides a superior "value proposition" to an increasing number of users, industries represented by complex systems and/or that require a critical mass of users/complementary products for growth to occur face additional challenges,[55] which may delay industry formation. Meeting these challenges might require agreements on standards, new methods of value capture and industry organization, government support for R&D, government purchases, new or modified regulations, new licenses, or even new ways of awarding licenses.

THINKING ABOUT THE FUTURE

Chapters 9 and 10 of Part IV use the conclusions from previous chapters to analyze the present and future of selected technologies. Chapter 9 looks at a broad number of electronics-related technologies such as displays, wireline and mobile phone telecommunication systems, the Internet and online services (including financial and educational services), and human-computer interfaces. Building from the notion of a technology paradigm, this chapter shows how improvements in specific components such as ICs have enabled new system-based discontinuities to become technically and economically feasible. More important, it shows how one can use an understanding of the technological trajectories in a system, or in key components of it, to analyze the timing of new discontinuities such as three-dimensional displays, cognitive radio in mobile phone systems, cloud/utility computing for the Internet, and gesture and neural-based human-computer interfaces.

Chapter 10 looks at three types of clean energy and how the four broad methods of achieving improvements in performance and costs can help us better analyze the potential for improvements in wind turbines, solar cells, and electric vehicles, so that we can provide better guidance on appropriate policies than can the typical emphasis on cumulative production. An emphasis on cumulative production says that the costs of clean energy fall as more wind turbines, solar cells, and electric vehicles are produced, that this "learning" primarily occurs within the final product's factory setting as automated equipment is introduced and organized into flow lines, that the extent of this learning depends on organizational factors, and that demand-based incentives are the best way for this learning to be achieved. Governments have responded to this emphasis on cumulative production by implementing demand-based subsidies, which firms have responded to by focusing on the production of existing technologies such as current wind turbine designs, crystalline silicon–based solar cells, and hybrid vehicles with existing lithium-ion batteries.

However, applying the three broad methods of achieving advances in performance and cost—notably, improvements in efficiency, geometrical scaling,[56] and key components—to clean energy leads to a different set of conclusions about policies. These policies involve the development of newer technologies and those that appear to have more potential for improvements than the ones currently emphasized.

For wind turbines, the key issue is geometrical scaling. Chapter 8 describes how costs per output have fallen as the physical length of turbine blades and towers have been increased, with these increases in scale requiring stronger and lighter materials. Thus, government policies should probably focus on the development of these materials through supply-based incentives such as R&D tax credits or direct funding of research. Furthermore, some evidence suggests that the limits to scaling have been reached with existing wind turbine designs, particularly those using existing materials, and so new designs are needed. Again, supply-based incentives such as R&D tax credits or direct funding of new designs will probably encourage manufacturers to develop new designs more than will demand-based subsidies.

For solar cells, improvements come from a combination of increases in efficiency and reductions in cost per area, where the latter are primarily driven by both reductions in the thicknesses of material and increases in the scale of production equipment (both are forms of geometrical scaling). The largest opportunities for these improvements are in new solar cell designs such as thin-film ones that are already cheaper on a cost-per-peak-watt basis than are crystalline silicon ones. Unfortunately, crystalline silicon ones are manufactured far more than are thin-film ones because turnkey factories are more available for their manufacture and thus firms can more easily obtain demand-based subsidies for them. Therefore, as with wind turbines, governments should probably focus more on supply-based incentives such as R&D tax credits or direct funding of new forms of solar cell to realize the necessary improvements in efficiency and reductions in material thicknesses that appear possible with thin film.

For electric vehicles, the key component is an energy storage device (e.g., a battery), and thus appropriate policies should focus on this device and not on the vehicle itself. Chapter 10 describes how improvements in lithium-ion batteries, which currently receive the most emphasis from vehicle manufacturers, are proceeding at a very slow pace, and notes that large improvements are not expected to emerge in spite of the fact that such improvements are needed before unsubsidized electric vehicles can become economically feasible. Therefore, to encourage firms to look at new forms of batteries (or other forms of energy storage device such as capacitors, flywheels,[57] or compressed air), again, governments should probably focus on supply-based incentives such as R&D tax credits or direct funding of new forms of energy storage devices.

WHO IS THIS BOOK FOR?

This book is for anyone interested in new industries and in the process of their formation, including R&D managers, high-tech marketing and business development managers, policy makers and analysts, professors, and employees of think tanks, governments, high-tech firms, and universities. The information it presents will help firms better understand when they should fund R&D or introduce new products that can be defined as a new industry. It will also help policy makers and analysts think about whether technologies have a large potential for improvement and how governments can promote the formation of industries that are based on these technologies. Furthermore, it will help uncover those technologies that have a potential for large improvements and thus a potential to become new industries, which is much more important than devising the correct policies for a given technology.

This book is particularly relevant for technologies in which the rates of improvement in performance and cost are large and thus the frequency of discontinuities is high. For firms involved with these technologies, understanding when technological discontinuities might emerge is a key issue because these discontinuities often lead to changes in market share and sometimes lead to incumbent failure. They may even lead to changes in shares at the country level. For example, the emergence of technological discontinuities impacted the rising (and falling) shares of U.S., Japanese, Korean, and Taiwanese firms in electronics industries in the second half of the 20th century.

Thus, the information this book offers can help firms, universities, and governments better understand when discontinuities might emerge. On the one hand, scientists such as Kaku, in *Physics of the Future* (2011), Stevenson, in *An Optimist's Tour of the Future* (2011), and Diamandis and Kotler, in *Abundance* (2012)[58] discuss the scientific and technical feasibility of different technologies. On the other hand, business professors discuss the strategic aspects of new technology in terms of, for example, a business model.[59] This book helps one understand when scientifically and technically feasible technologies might become economically feasible and thus when firms, universities, and governments should begin developing business models and appropriate policies for them.

This book is also for young people, who have more at stake in the future than anyone else, and it has been written to help them think about their future. It will encourage students to think about where opportunities may emerge and thus the technologies they should study and the industries where they should begin their careers. In terms of opportunities, while the conventional wisdom is to focus students on customer needs or on what is scientifically or technically feasible, it is also important to help them understand those technologies that are undergoing improvements and how these improvements are creating opportunities in higher-level systems. This is something that few engineering classes do, partly because they focus heavily on mathematics (and are criticized for this).[60]

For example, helping students (and firms and governments) understand how reductions in the feature sizes of ICs, including bioelectronic ICs and MEMS (microelectronic mechanical systems), can help them search for new opportunities. My students have used such information to analyze 3D holograms, 3D displays, MEMS for ink-jet printing, membranes, wireless charging, wave energy, pico-projectors, 3D printing, different types of solar cells and wind turbines, cognitive radio, and new forms of human-computer interfaces (e.g., voice, gesture, neural), as well as to analyze the opportunities that are emerging from these technologies;[61] some of these presentations are a source of data in Chapter 9. Among other things, the final chapter discusses how this book can be used in university courses to help students think about and analyze the future.

Finally, the ideas discussed here can helps students and other young people look for solutions to global problems that will not be easily found. Without an understanding of technology change, how can we expect students to propose and analyze reasonable solutions? To put it bluntly, discussions of policies, business models, and social entrepreneurship are necessary but insufficient. New technologies and improvements in existing ones provide tools that our world can use to address global problems. Therefore, proposed solutions should consider the potential for and rate of technology improvements. For example, Chapter 10 analyzes three types of clean energy and concludes that, because the potential for improvements is mixed, more radical solutions are probably necessary. We need to ask students the right questions and give them the proper tools so that they can do this type of analysis and propose more radical solutions.

1
What Determines the Potential for New Technologies and Thus New Industries?

Understanding the potential for new technologies is highly problematic. Consider, for example, nuclear power. On September 16, 1954, Lewis Strauss, then chairman of the U.S. Atomic Energy Commission, in a speech to the National Association of Science Writers, said, "Our children will enjoy in their homes electrical energy too cheap to meter." Although there is some disagreement about whether he was referring to nuclear fission or nuclear fusion,[1] children born in the 1950s still do not have access to energy that is too cheap to meter, and it is highly unlikely that those born in the 21st century will have it. In fact, the opposite may be true.

Similar examples can be found in a book written in 1967 by Herman Kahn, considered one of the greatest visionaries of his time, and Anthony Weiner. Kahn and Wiener described technologies they believed would be widely used by the year 2000. In 2002, Richard Albright created a panel of experts to assess these forecasts. It was concluded that fewer than 50 percent of the forecasted innovations had occurred before the end of the 20th century. By sector, a larger percentage were correct in computers and communication (80 percent) than in aerospace (20 percent) and infrastructure and transportation (30 percent). Consistent with this book's analysis, Albright concluded that the greater accuracy in forecasts for computers and communication was the result of greater improvements in the underlying technologies for them, such as integrated circuits (ICs), magnetic storage, and optical fiber.[2] On the other hand, the lower forecasting accuracy for aerospace was the result of exaggerated hype generated by the Apollo program. Exaggerated hype may now be occurring with electric vehicles and other clean energies.

Part I describes the concepts of technology paradigms and geometrical scaling and how they can be used to better assess the potential of new technologies. Chapter 2 provides an overview of technology paradigms, while Chapter 3 focuses on a key aspect, geometrical scaling. Although any method is prone to exaggeration and hype such as occurred in some of Kahn and Wiener's examples, using a technology paradigm requires gathering performance and cost/price data and considering both the components that make up a technological system and the concepts that form the basis of the technology whose potential is being assessed.

For example, if Lewis Strauss had analyzed the technology paradigm for nuclear power in 1954, he would have noted several key things: (1) that boilers, turbines, and generators are needed for nuclear fission just as they are for fossil-fired power; (2) that the benefits from increasing the scale of boilers, turbines, and generator technologies had almost been attained by the 1950s; (3) that, although nuclear fuel has higher energy densities than do other fuels, its costs of extraction, shipping, and processing on a weight or volume basis are also higher, for safety reasons; and (4) that the costs of plant construction are higher for nuclear fuel, also for safety reasons. These facts should have alerted him to the nuclear energy's potential limitations.

2

Technology Paradigm

This chapter summarizes and contrasts the paradigms for six basic technologies and more than 35 "subtechnologies" (see the Appendix for methodology). Building from Giovanni Dosi's characterization of a technology paradigm, it does this in terms of (1) a technology's basic concepts or principles and the trade-offs that are defined by them; (2) the directions of advance within these trade-offs, which are defined by one or more technological trajectories; (3) the potential limits to these trajectories and their paradigms; and (4) the roles of components and scientific knowledge in these limits.[1] The concepts or principles that form the basis of a technology define the trade-offs between cost and various dimensions of performance, and their characterization and implementation are supported by advances in science. This book distinguishes between a physical phenomenon, which exists independently of humans, and the use of a concept or principle to exploit it for a specific purpose.[2] Advances in our understanding of a physical phenomenon form a base of knowledge that helps us find new concepts or principles, and these in turn help us find better product and process designs.

This chapter primarily focuses on the second and third items listed above and shows that there are four main methods of achieving advances along a technological trajectory: (1) improving the efficiency with which basic concepts and their underlying physical phenomena are exploited; (2) radical new processes; (3) geometrical scaling; and (4) improvements in "key" components.[3] Since Chapter 3 focuses on scaling and radical process improvements as part of scaling in production equipment, this chapter focuses on the other two methods, the first of which includes finding materials that better exploit basic concepts and their underlying physical phenomena. Finding better materials (and radical new processes, which are not discussed in this chapter) is particularly important for the low levels in a nested hierarchy of electronic subsystems, while improvements in "key" components are more important for the higher levels.

A major goal of this and other chapters is to help managers, engineers, professors, and students better analyze the potential of a new technology. Understanding the extent to which improvements have occurred in old and new technologies can help us determine the extent to which they might occur in the future and thus when a new technology might provide a superior "value proposition" to an increasing number of users. In particular, since this chapter shows how improvements in performance and cost/price in a technological trajectory generally occur in a rather smooth and incremental manner over multiple generations of discontinuities, and shows that dramatic improvements do not occur following the emergence of a technological discontinuity, an understanding of these incremental improvements enables one to roughly understand the near-term trends in performance and cost/price for both a system and its components.

ENGINES AND TRANSPORTATION TECHNOLOGIES[4]

Table 2.1 summarizes the technology paradigms for three types of engine. In a steam engine, steam does work by pushing out a piston while a vacuum (from the condensing steam) pulls the piston back in. Often defined as the key technology in the industrial revolution, the first steam engines had efficiencies as low as 0.5 percent (see Figure 2.1), and, even with the addition of a separate condenser by James Watt in 1765, efficiencies were still just a few percent.[5] They increased only as better components, tighter tolerances, and better controls emerged and as the perceived usefulness of steam engines stimulated research on thermodynamics, combustion, fluid flow, and heat transfer. Furthermore, as described in Chapter 3, increases in scale led to higher efficiencies through a positive interaction between larger scale, higher pressures, and higher temperatures.[6]

Steam engines revolutionized mining, factories, and later transportation. Steam-powered pumps allowed deeper mines to be dug and larger factories to be built, the latter enabling some of the increases in equipment and factory scale that are described in the next chapter. Because of more than 100 years of improvements in both their efficiencies and scale, locomotives and steamships became possible in the 19th century (see Table 2.2). Without high efficiencies, a low ratio of power to weight made it difficult to move an engine, much less cargo or passengers. For land transportation, heavy weight meant that engines

TABLE 2.1

Technology paradigms for engine technologies

Type of engine	Basic concept/principle	Basic methods of improvement
Steam engine (early 1700s)	Power is generated and work done by pressurized steam pushing against a piston	Increased efficiency from higher temperatures, pressures, and size (geometrical scaling) and from better controls over fuel, air, and heat
ICE (mid-1800s)	Power is generated and work done by an explosion and subsequent expansion of gaseous fuel pushing against a piston	
Jet engine (mid-1900s)	Combustion of high-temperature and high-pressure fuel provides thrust	

NOTE: ICE, internal combustion engine.

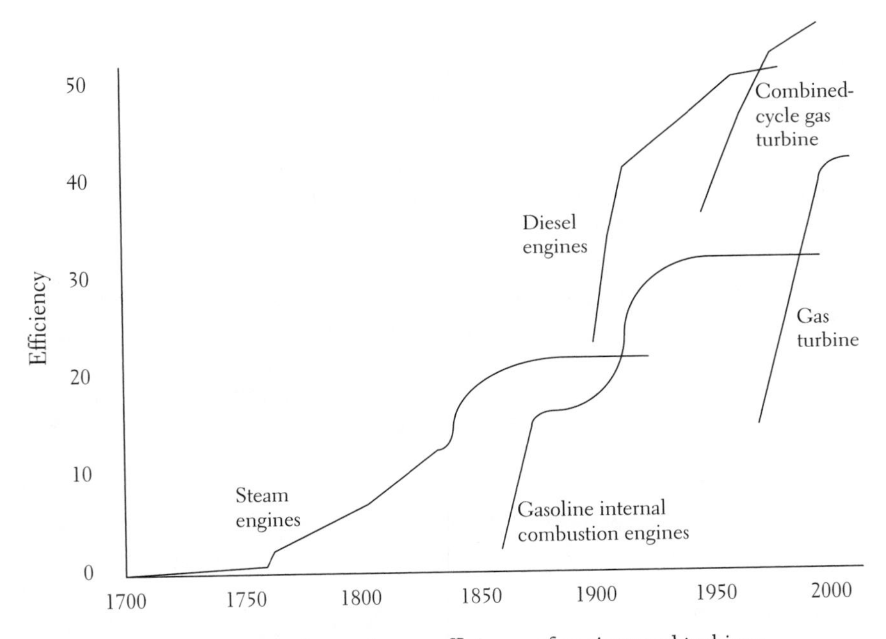

FIGURE 2.1 Improvements in maximum efficiency of engines and turbines

SOURCE: Republished with permission of ABC-CLIO, Inc., from *Energy Transitions: History, Requirements, Prospects*, Vaclav Smil, © 2010; permission conveyed through Copyright Clearance Center, Inc.

TABLE 2.2

Technology paradigms for transportation technologies

Technology	Basic concept/principle	Basic methods of improvement
Locomotive	Output from steam engine turns wheels; wheels run on track	Geometrical scaling Aerodynamic designs Lighter materials
Steamship	Output from steam engine (later ICE) turns propeller	
Electric train	Electricity powers rotation of wheels	
Automobile	Output from ICE turns wheels; wheels move over ground	
Aircraft	Pushed forward by output from ICE (later by jet engine); wings provide "lift"	

SOURCE: Adapted from Smil (2010, fig. 1.2).
NOTE: ICE, internal combustion engine.

were first used on heavy rails and not roads. For water transportation, initially poor efficiencies (and small scale) caused them to be used on rivers long before they were used on oceans.

Road and air transportation required a much lighter and smaller engine and thus one with a new form of technology paradigm. The internal combustion engine (ICE) uses a spark to ignite a small explosion and the subsequent expansion of gaseous fuel to push a piston. The explosion enables the ICE to generate much more power per weight or volume than a steam engine can. However, it required a "spark," which could not be achieved until electric batteries (see the next section) existed. Other necessities included a better understanding of combustion, which partly came about through Antoine Lavoisier's separation of oxygen from other gases; a plentiful supply of highly volatile fuel such as gasoline; and improvements in materials and equipment for forming and cutting metal. The latter improvements had only emerged by the second half of the 19th century.[7]

Like the steam engine, the ICE also benefited from improvements in efficiency and increases in scale, and these improvements, along with a better understanding of the aerodynamics of lift, made human flight possible. According to Bernoulli's principle, first described in his book *Hydrodynamica* in 1738 and applied to flight by his successors, when air moves faster over the top than under the bottom of a wing, there is lower pressure above than below and thus the wing experiences "lift."[8] The challenge was to propel an engine and a human at a high enough speed for them to experience lift.

The necessary improvements in the efficiency and weight of ICEs in the late 1800s were driven partly by the market for automobiles and partly by improvements in a wide variety of manufacturing processes in diverse industries. These improvements in weight and efficiency continued throughout the 20th century and resulted in a 300-fold reduction in the mass-to-power ratio of ICEs.[9] Nevertheless, there are limits to the ICE paradigm. Pistons can only move so fast, and the efficiency of propellers declines as the speed of sound is approached or as altitude is increased.

Jet engines involve a different technology paradigm that permits aircraft to fly at high altitudes where air density and thus friction are low. Like the rockets first used in the 15th century, they depend on Newton's Law of Action and Reaction: the exhaust from the combustion of high-temperature, high-pressure fuel propels an airplane forward. Realizing this technology paradigm and advancing it required compressors, turbines to drive the compressors, and materials that could withstand high temperatures. Most of these "components" did not become available until the mid-20th century, when some of them were borrowed from the electric power industry (e.g., turbines). These improvements, along with increases in the scale of engines and aircraft (addressed in the next chapter), led to increases in temperature and pressure and thus increases in engine efficiency of about 30 percent between 1960 and 2000.[10] Now it is the exhaust from the jet and the ICE in the form of carbon dioxide that is driving a search for a new technology paradigm for engines.

ELECTRICITY GENERATION[11]

Table 2.3 summarizes the technology paradigms for three types of electricity generation and several subtypes within the most common type of "generators and turbines." Building from Luigi Galvani's research on the movements of dead frogs, Alessandro Volta built the first battery in 1800. However, it was Michael Faraday who explained in the 1820s that electricity was not inexhaustible and that it came from chemical reactions within the frog and battery. Faraday's scientific explanation of electricity and its interaction with magnetism, along with improvements in batteries and other "components," made technologies such as telegraphs and electrolysis possible.

Electrolysis allowed scientists to isolate elements that do not appear as elements in nature. One reason for this is that these elements are very reactive, and it is highly reactive elements that make good batteries, since the difference in reactance between the anode and the cathode in a battery largely determines energy and power densities. Batteries with higher energy and power densities store more energy per weight or volume and provide more power per weight or volume, respectively, than do those with lower energy and power densities. They also often have lower costs per unit of energy, since battery cost is often a function of volume or weight (as is true for other

TABLE 2.3

Technology paradigms for electricity generation technologies

Technology	Basic concept/principle	Basic methods of improvement
Battery	Transforms chemical energy into electrical energy	More reactive, higher current-carrying, lighter materials
Generator/turbine	Movement of loop of wire between poles of magnet by turbine generates electricity Turbine rotation driven by water, wind, or steam; steam generated by variety of sources	Higher temperature, pressure, and scale Higher energy density of fuels (e.g., nuclear)
Photovoltaics	Absorption of photon releases energy equal to "band gap" of material	Materials with higher efficiencies even when very thin Larger-scale production equipment

SOURCE: "Multi-Year Program Plan, FY '09-15," Solid-State Lighting Research and Development, March 2009, cited in Williamson (2010) and Koh and Magee (2008, fig. 1.a).

technologies[12]). Thus, the history of batteries has been a search for materials with high reactivity for the cathode and low reactivity for the anode, as well as for higher current-carrying capacity, low weight, and ease of processing. The creation of the periodic table by Dmitri Mendeleyev in 1869 and the gradual accumulation of knowledge in material science in the 19th and 20th centuries also helped scientists search for new and better materials, and new materials continue to be found.

Improvements in the last few decades have come from the use of completely new materials such as lithium and small changes in the particular combination of lithium and other materials (see Figure 2.2[13]). This has led to a doubling of energy densities for Li-ion batteries in the last 15 years, and some observers expect a similar doubling to occur in the next 15 years from modified forms of lithium such as Li-air.[14] Chapter 10 addresses the role of batteries and other energy storage technologies in electric vehicles.

Most electricity is of course generated by the movement of a turbine in an electricity-generating station. The rotation of a turbine can be driven by wind, water, or steam, with steam the most common medium. Similar to steam engines, steam turbines are primarily powered by the burning of oil, coal, or gas, but the steam can also come from sources such as nuclear fission, geothermal, or solar thermal. The rotation of the turbine moves an electrical conductor perpendicular to a magnetic field, which operates on the principles identified by Michael Faraday in the 1830s. The implementation of these turbines for generating electricity required improvements in a large number of complementary technologies such as alternators, rotors, dynamos, and transmission

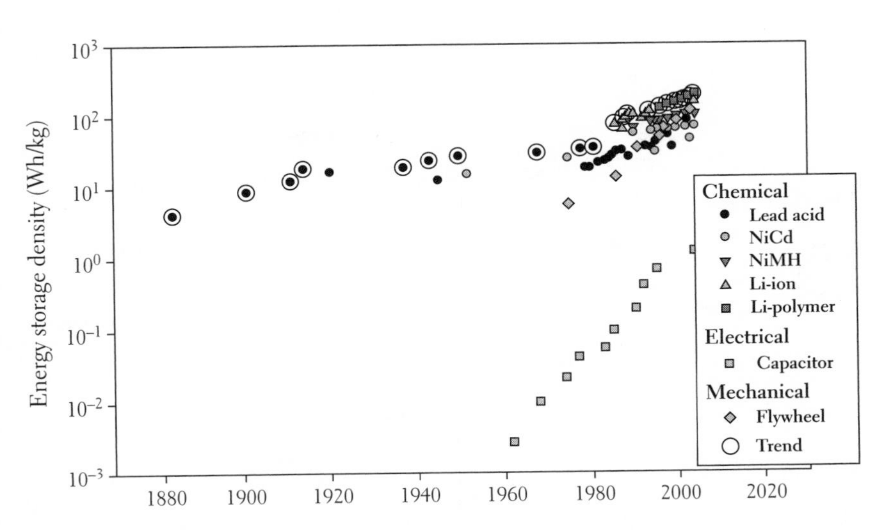

FIGURE 2.2 Improvements in energy storage density
SOURCE: Koh and Magee (2008).

lines, many of which depended on improvements in manufacturing processes for mechanical and other components during the 1800s. Implementation also required the development of a major application for electricity, which turned out to be electric lighting, and this depended on the development of an incandescent lightbulb (see the next section). Since electricity's introduction in the 1870s, its price has dropped dramatically because of improvements in efficiency and increases in scale, which are discussed more in the next chapter.

Another aspect of a technology paradigm for most conventional electricity-generating stations is the production of carbon dioxide, particulates, sulfur and nitrous oxides, and other pollutants. These are by-products of burning coal, oil, and gas, and the difficulties in capturing them, particularly carbon dioxide, have started a search for new sources of electricity such as wind and solar that do not emit carbon dioxide and other environmental pollutants. Wind turbines and solar cells are analyzed in Chapter 10.

LIGHTING AND DISPLAY[15]

Table 2.4 summarizes the technology paradigms for several types of lighting and display. These technologies required advances in the science of electricity and electric discharge tubes; many of those in the lower half of the table also depended on advances in the science of crystals and semiconductors.

TABLE 2.4

Technology paradigms for lighting and display technologies

Technology	Basic concept/principle	Basic methods of improvement
Electric arc light	Passing a current across two electrodes generates heat/light	Materials and gases with high luminosity/input power ratio
Electric discharge tube	Voltage difference across two electrodes or across filament connecting two electrodes in a vacuum causes emission of:	
Incandescent lights	visible light (as filaments become hot and incandesce)	Filaments with high luminosity/input power ratio
CRT	electrons from one electrode where electrons striking phosphors cause photon emission[a]	Cathodes that efficiently produce electrons; phosphors that better luminesce
Fluorescent light	ultraviolet light—high-energy photons cause emission of visible light when they strike phosphors	Gases that efficiently emit ultraviolet light; phosphors that better fluoresce
Other lights	visible light in gases (e.g., mercury, sodium vapor)	Gases with high luminosity/input power ratio
LED	Semiconductor diode emits light when voltage is applied	Semiconductors with high luminosity/input ratio and lasers with high coherence; reducing size can reduce cost
Semiconductor laser	Semiconductor diode emits "coherent" light in single wavelength	
LCD	Alignment of crystals modulates external light source (e.g., backlight); alignment typically depends on input voltage:	
Passive-matrix	pixel output depends on voltage to row and column via multiplexing	Increase resolution with more pixels (improvements limited by need to multiplex)
Active-matrix	pixel output depends on voltage applied to each pixel with transistor	Increase transistor density to improve resolution, viewing angle; thinner materials lead to lower cost
OLED	Organic materials emit light depending on input voltage and material band gap	Materials with high luminosity/input power ratio; thinner layers (to reduce costs)

SOURCE: Agilent Technologies and Misra et al. (2006).

NOTE: CRT, cathode ray tube; LCD, liquid crystal display; OLED, organic light-emitting diode.

[a]The direction of the electrodes can be controlled so that the electrons hit certain phosphors.

Advances in electric discharge tubes were first made in the second half of the 19th century by Heinrich Geissler and William Crookes, whose tubes still bear their names, and further advances were helped by the basic research of J. J. Thompson and Albert Einstein, among others.[16]

As with batteries, with electric discharge tubes there has been a search for materials that more efficiently exploit basic concepts and underlying physical phenomena. This search, helped by the accumulation of knowledge in various material sciences, has been for materials and gases that convert a larger percentage of electricity into visible light when a voltage is applied across two electrodes or across a filament connecting two electrodes in a vacuum.

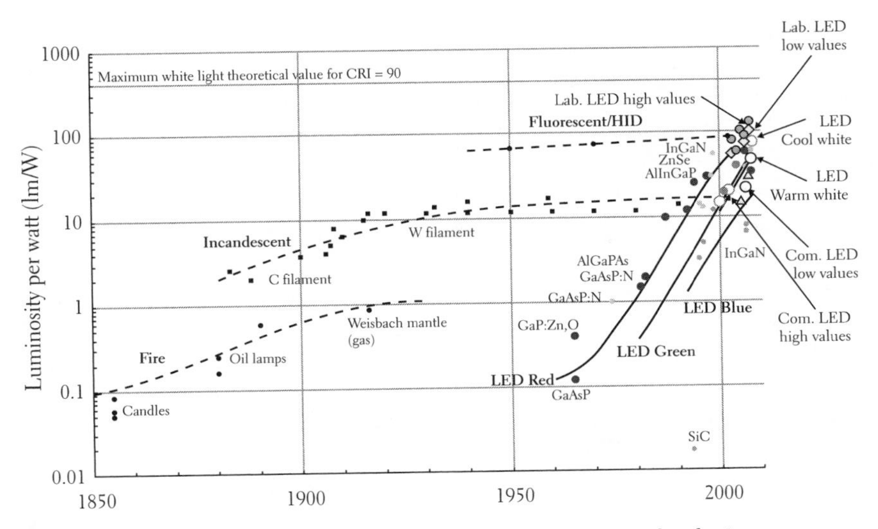

FIGURE 2.3 Improvements in luminosity per watt for various technologies
SOURCE: © 2009 IEEE. Reprinted, with permission, from Azevedo et al., *Proceedings of the IEEE* 97 (2009): 481–510.

For incandescent lights, the search has been for filaments that translate more electricity into visible light than into heat, but it has made little progress since the 1930s (see Figure 2.3). Engineers and scientists have had more luck with fluorescent lights. Although the gases used in these lights mostly emit ultraviolet light, so-called phosphors mostly emit visible light upon absorbing ultraviolet light, which enabled higher efficiencies than for incandescent lights. For other technologies, the search has been for gases that emit visible light when a voltage is applied across two electrodes that encompass them. For all of these lights, advances in pumps and seals (mostly spin-offs from steam engines) were needed to create a vacuum; these advances (along with advances in glass tubes) were realized in the second half of the 19th century. In combination with rising wages, these improvements led to a dramatic drop in the hours worked needed to acquire a lumen of lighting.[17]

Nevertheless, limits to the technology paradigms for lighting based on electric discharge tubes emerged many years ago. Although there is currently an emphasis on replacing incandescent bulbs with so-called compact fluorescent tubes, limited improvements in efficiency or reductions in size (and thus costs) suggest that these tubes are a temporary solution before something else replaces them. For example, LEDs emit light when a voltage is applied across a semiconductor diode, which is based on the phenomenon of electroluminescence. Semiconductor lasers take this one step further and emit light for a single wavelength and phase. The phenomenon of electroluminescence was first identified in the late 19th century, and the first semiconductor LEDs and lasers were independently produced in 1962 by Robert Hall and Nick Holonyak, respectively.

Once the first LEDs and lasers had been constructed, the challenge was to find semiconducting materials and device structures with a high ratio of luminosity to power consumption (i.e., watt) for LEDs (see Figures 2.3 and 2.4), a high coherence (i.e., low dispersion) for lasers, and low costs for both, which depended on the accumulation of knowledge for semiconductors.[18] LED costs fell by more than three orders of magnitude between 1975 and 2005 partly from making them on larger wafers (i.e., scaling), making them with smaller packages, and introducing radical new processes. For LEDs, the U.S. Department of Energy concluded that there is still a large potential for improvements in efficiency and costs, and thus they will likely diffuse at some time in the future.[19]

An organic LED (OLED) might also be used for displays in place of cathode ray tubes (CRTs) or liquid crystal displays (LCDs). A CRT is a form of

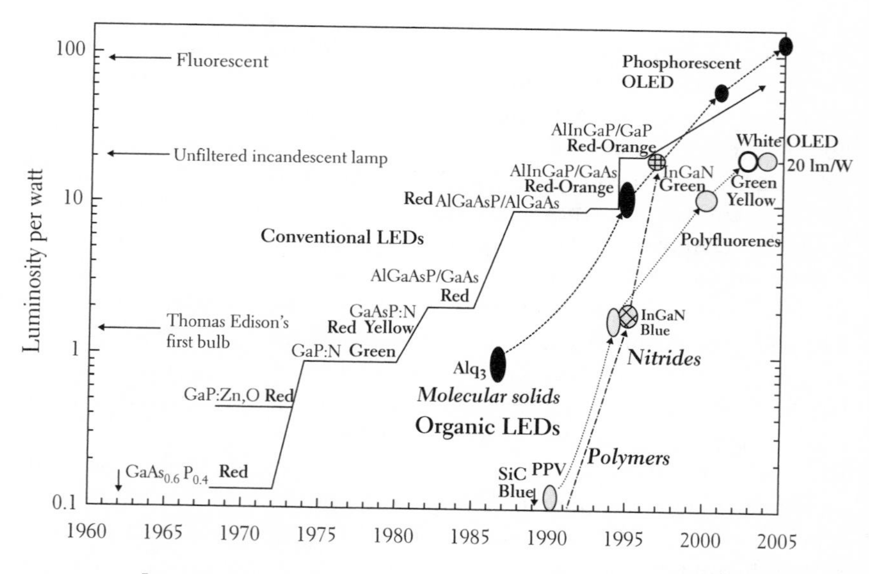

FIGURE 2.4 Improvements in luminosity per watt for LEDs and OLEDs

SOURCE: From Sheats et al., "Organic Electroluminescent Devices," *Science* 273, no. 5277 (1996): 884–888. Reprinted with permission from AAAS. Adaptation reprinted with permission from Changhee Lee.

NOTE: PPV, polyphenylene vinylene.

electric discharge tube that was initially used as an oscilloscope but is most well known for its long life as a display in television receivers. In a CRT, one electrode emits electrons that strike phosphors, causing the phosphors to luminesce.[20] By controlling the direction of the electrons with an electric field, one can determine the specific phosphors on a glass tube that will be struck. Using three electrodes or so-called electron guns and the right type of phosphors, color images can be displayed on the television screen. Like other forms of electric discharge tube, many improvements in CRTs came from finding better phosphors and better materials for the electron gun.

Limits to the technology paradigms for cathode ray tubes began to emerge many years ago. Like incandescent lights, CRTs' miniaturization (and thus cost reduction) has been severely constrained by the size of electrodes, glass bulbs, and sockets, and their resolution is constrained by similar problems (difficulties in controlling electrodes, emitted electrons, and impacted phosphors). As for LCDs, although the properties of liquid crystals had been identified by the late 19th century, it was not until scientists were able to control them with an electric field in the mid-1960s that interest in them grew. Applying an electric field causes LCDs to align in the appropriate direction and thus either block or transmit polarized light from an external source such as a so-called backlight. Different electric fields applied to different regions, or "pixels," in a liquid crystal cause light to be either passed or blocked by the different pixels, forming an image on a display. As with many of the previous examples, finding the appropriate liquid crystal materials and materials for polarizers and color filters took many years of research in the 1960s, 1970s, and 1980s. These advances depended on the accumulation of scientific knowledge in crystalline materials and the ability to use semiconductor-manufacturing equipment to deposit and form patterns in them. Semiconductor-manufacturing equipment also facilitated a change from so-called passive-matrix to active-matrix LCDs,[21] whose cost reductions are now largely driven by increasing the scale of this equipment and reducing the thickness of the materials.[22]

A bottleneck in further miniaturizing LCDs has been the size of an external light source such as a backlight. One way this problem is being addressed is by using LEDs in place of so-called cold cathode fluorescent lights since they can be made much smaller. Another option is to replace the entire LCD with LEDs. Unfortunately, different-color LEDs require different combinations of rather esoteric semiconductor materials, which are currently difficult to combine on a single substrate.

With OLEDs, depositing different polymers, each emitting a different color, on a single substrate (usually glass) using ink-jet printing is much easier than depositing different semiconductor materials, each emitting a different color, on a semiconductor substrate. While LEDs and semiconductors in general require very complex deposition processes at high temperatures, different polymers, including those for different colors, can be roll-printed onto a very thin glass substrate at room temperature using conventional printing techniques. This enables OLEDs to be potentially thinner, more flexible, and cheaper than LCDs, while the elimination of an external light source enables them to use less power and have higher viewing angles. As with many of the previous examples, following the identification of polymers in the early 1950s by A. Bernanose and the creation of the first diode from them in the 1980s, finding the appropriate polymers (see Figure 2.4) has taken many years and has required many advances in conductive polymer science. These polymers must emit light in the three primary colors (red, blue, and green), they must do so for a high luminosity per watt, and they must have a long life. Currently, one major limitation of polymers is their shorter lifetimes than those of LCDs, so there is still a search for better materials and a need for a better understanding of their science. Again, as with many of the examples previously given, the key question is whether these materials will be found. Many material scientists believe they will.[23]

INFORMATION TECHNOLOGY[24]

Table 2.5 summarizes the technology paradigms for several types of information technology (IT), some of which were mentioned before. The first technology with the capability to quickly process information was the vacuum tube, which is a type of electric discharge tube that was first implemented in the form of a diode in 1905. Unlike that of mechanical devices, the speed of a vacuum tube is limited by the speed of electrons in filaments (and other wires) that connect multiple electrodes. Increases in the number of electrodes expanded the functions that a vacuum tube could perform. Terms such as *triode*, *tetrode*, and *pentode* signify the number of electrodes. In a three-terminal device (tetrode), which is the most widely used vacuum tube, one input amplifies, modifies, or switches a second input, resulting in an output that depends on the value of both inputs. The size of the tubes and the heat loss of filaments were both gradually reduced, and there was a search for better materials for filaments and electrodes, just as there was for the incandescent lightbulb. Nevertheless, it has been easier to reduce the sizes and increase the speeds of transistors than the sizes and speeds of vacuum tubes.

Transistors are also three-terminal devices, but, like integrated circuits (ICs), they rely on a technology paradigm different from that for vacuum tubes. The input and output terminals of a transistor are formed on a thin

TABLE 2.5

Technology paradigms for information technologies

Technology	Basic concept/principle	Basic methods of improvement
Three-terminal device	Amplify, switch, or modify electrical signal by controlling movements of electrons between:	
Vacuum tube	electrodes in vacuum	Increase number of electrodes; reduce size/heat loss of filaments
Transistor (alone or in IC)	source and drain in thin layer of semiconductor material	Use better materials; reduce distance between source and drain; reduce material thickness to increase speed and reduce cost
Memory storage	Save 1s and 0s in a memory cell where value depends on:	
Memory cell in IC	output voltage	Reduce cell size to increase speed and density/reduce cost
Magnetic	magnetization of region	Reduce region size to reduce cost/increase speed and density
Optical	existence of pit in metal disk formed by semiconductor laser and read by reflections to photocell	Reduce pit size with smaller wavelengths to reduce cost/increase density and speed
Computer	Execute stored programs	Largest benefits from better components

NOTE: IC, integrated circuit.

substrate of semiconductor material, which has enabled dramatic reductions in transistor size, power consumption, and cost. Two key events in the development of this technology paradigm were the creation of the first junction transistor in 1948 by John Bardeen, Walter Brattain, and William Shockley, and the combining of multiple transistors, resistors, and capacitors on a single IC in 1959, accomplished independently by Jack Kilby and Robert Noyce. Between these events was a successful search for better materials and better transistor and process designs that simultaneously involved the rapid accumulation of scientific knowledge of semiconductor materials. Subsequent efforts focused on reducing the distances between transistor inputs and outputs, the width of metal lines connecting transistors, and the thickness of various layers. These efforts led to dramatic increases in the number of transistors (as well as memory cells) that could be placed on an IC. As described in Chapters 3 and 6, reducing an IC's dimensions has created exponential improvements in its cost and performance.

Both vacuum tubes and ICs enabled the introduction of and improvements in many electronic products, some of which are described elsewhere in this and other chapters. Vacuum tubes were widely used in early radios, televisions, telephone switches, wireless systems, and computers, and ICs have driven improvements in these products. Improvements of ten orders of magnitude in

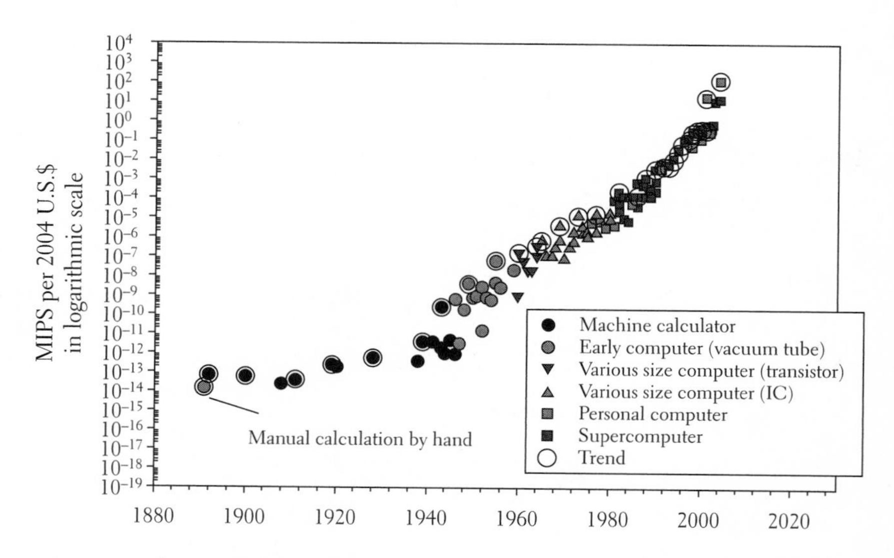

FIGURE 2.5 Improvements in MIPS (million instructions per second) per price level (in U.S. dollars)
SOURCE: Koh and Magee (2006).

ICs have led to equally large improvements in various electronic products (for computers, see Figure 2.5[25]), which are analyzed more in Chapters 4 and 6. In spite of improvements, however, the dependence of these products on ICs and other electronic components means that typically less than 5 percent of their costs involve assembly; the rest involve ICs, displays, other electronic components, and magnetic storage.[26]

Information can also be stored using magnetism. On the basis of Michael Faraday's work on electricity and magnetism in the 1820s and 1830s, a material can be magnetized by varying the current in a local read-write head. However, using Faraday's principles to design a storage system required improvements in many components. The first magnetic storage systems used so-called magnetic cores, which were small magnets connected by wires. Improvements in manufacturing processes later enabled small magnetic cells to be formed on thin layers of magnetic material that were deposited on drums, hard disks, and floppy disks. These improvements were supported by an expanding base of knowledge in magnetic materials and sputtering equipment. As discussed in the next chapter, improved sputtering equipment enabled reductions in the size of these cells and these reductions led to exponential improvements in the recording density of magnetic platters and tape and thus in the cost and performance of magnetic storage systems.

Optical storage relies on small wavelengths of light to alter the reflectivity of a small region on a metal disk. As with optical fiber, semiconductor lasers and diodes respectively produce and sense light. In the case of optical discs, a photodiode senses whether light has been reflected by a small region on a metal disk. Since the use of smaller wavelengths of light reduces the size of this region and thus the storage cell, there has been a search for lasing materials with smaller wavelength. Unfortunately, this search is limited by the wavelength of light and the cost of smaller-wavelength light sources, and so fewer improvements have been made here than in ICs and magnetic disks. This limitation, in combination with the shape of an optical disc, the size of the read-write head, and the rapid increases in video downloading over the Internet, may mean that there will be few additional improvements in storage density.

Moreover, further reductions in IC feature sizes are becoming more difficult (see Chapters 3 and 6), and thus ICs may be replaced by new technologies such as phase change memory, magnetic random-access memory, photonics (optical computing), and organic transistors. Since this chapter has already described one application of organic materials (OLEDs) and Chapter 8 discusses a second one (solar cells), it is useful to mention a third application, organic transistors. These are already used as display drivers on OLEDs, and they could be used to produce new forms of ICs and displays. Although their use in place of ICs has been limited by their low mobility (which impacts speed), improvements of eight orders of magnitude were made between 1982 and 2002 for polycrystalline transistors; similar levels of improvement for single-crystalline transistors were achieved between 1997 and 2007.[27] Can another two orders of magnitude improvement be made so that OLEDs' mobility will rival that of silicon transistors?

TELECOMMUNICATION[28]

Table 2.6 summarizes the technology paradigms for several types of telecommunication technologies. A telegraph transmits a series of dots and dashes from a sender to a receiver through a metal line by controlling the time the line is "open." Based on Faraday's Law, an open line on the sender's side was sensed on the receiver's side by whether a magnetic compass was deflected. Telephones depended on both Faraday's Law and Ohm's Law; the latter was published by Georg Ohm in 1827. In the first telephones, sound waves were converted into electrical signals using a vibrating diaphragm that was

TABLE 2.6

Technology paradigms for telecommunication technologies

Technology	Basic concept/principle	Basic methods of improvement
Telegraph	Opening a circuit deflects compass needle; status of needle transmitted over metal wires	Multiplexing, automation
Telephone	Sound waves converted to electrical signals with microphone	Multiplexing (transmit multiple signals in parallel)
Copper line	Signal transmitted in voltage waves	Increase capacity with cladding (e.g., coaxial cable to reduce signal/power loss), digitalization, higher frequencies
Optical fiber	Optical signal generated by semiconductor lasers, transmitted in light waves, converted to electrical signals by photocells	Increase capacity with purer fiber (to reduce losses), lower dispersion, and shorter-wavelength lasers
Wireless	Wireless transmission of voice and data signals by modulation of electromagnetic waves	Reuse frequency spectrum in more cells; digitalization of signal and better algorithms enable more efficient use of frequency spectrum
Internet	Data transmitted in packets "routed" through various lines, and reassembled at end points	Increase capacity with better "components" and more efficient management of packets

SOURCE: Adapted from "Trends Report" from 2010 International Solid-State Circuits Conference.

connected by two metal plates separated by carbon granules. Vibrations in the diaphragm changed the resistance in the carbon granules and thus changed the current levels in a copper line. Adequate diaphragms and carbon granules were not available until the second half of the 1800s.

The basic limitations for the paradigm of copper lines were their installation cost and their low capacities, as well as the high cost of manual switching. The first two problems were partially addressed by enabling multiple messages and later calls to be made simultaneously through a technique called multiplexing. The third was addressed by electrical switches. The implementation of switches and multiplexing on a large scale was made possible by improvements in electronic components such as vacuum tubes, transistors, and ICs.

However, by the time adequate ICs were available in the 1970s, optical fiber was emerging, which has a much higher potential capacity than do copper wires. Although the concept of optical fiber was partially demonstrated in the 1840s by Jean-Daniel Colladon and Jacques Babinet (it had been predicted by Newton's Law of Refraction), it was considered impractical until two advances occurred in the 1960s. First, the semiconductor laser was introduced; second, Charles Kao and George Hockham recognized that attenuation in fibers was caused by impurities and that these impurities could be removed.

Both advances were supported by an expanding base of knowledge in material science. Along with improvements in the performance of lasers and IC-based amplifiers, these events motivated improvements in the purity of glass, and by the 1980s copper lines were being replaced by optical fiber on a large scale both in main trunk lines and in local area networks (LANs).[29]

The technology paradigm for wireless voice communication is similar to and builds on the science behind broadcast radio. The first wireless systems were introduced in the 1920s for police and military. Large transmitters and receivers sent and received voice signals; a single transmitter covered a large area and a receiver filled the trunk of a car. Dividing a large area into small cells enabled the same frequency spectrum to be reused in each cell and thus greatly expanded the capacity of wireless systems. This concept and its underlying science were developed at Bell Labs in the 1940s, and so-called analog cellular systems had been introduced in many countries by the early 1980s. Modern systems use smaller cells or increasingly sophisticated algorithms to convert audio signals to digital signals and back; in doing so they increase their capacity and that of specific frequency bands (i.e., efficiency). All of these system-level improvements required improvements in electronic components such as ICs.

Packet communication, which can be applied to either wireline or wireless systems, involves a different technology paradigm from that for a circuit-switched system. Instead of dedicating a line to a voice call or a data transmission in either a wireline or a wireless system, a voice call or data transmission is divided into packets and these packets are "routed" through various lines and reassembled at end points. Packet technology was first conceived by Leonard Kleinrock in the late 1950s, and improvements in the capacity of packet systems came from improvements in glass fiber, ICs, lasers, photosensors, and other electronic components in a fiber-optic system.[30] As shown in Figure 2.6, improvements in data rates for wireline (shown for Ethernet LANs and USB storage devices) preceded those for wireless packet systems such as wireless LAN and WPANs (wireless personal area networks).

Theoretically, the limits to improvements in wireline and wireless telecommunication systems are determined by Shannon's Law, which states that the information capacity (C) of a system in bits per second is determined by the bandwidth (B) and the signal-to-noise ratio (S/N) in the following equation: $C = B * \log(1 + S/N)$. Increasing the bandwidth and the signal-to-noise ratio requires improvements in IC processing power and, for wireline systems, improvements in fiber-optic cable. As long as these improvements continue to be made, further improvements in telecommunication systems are likely to occur. For example, in theory gamma rays, which oscillate at 10^{24} Hz, can be

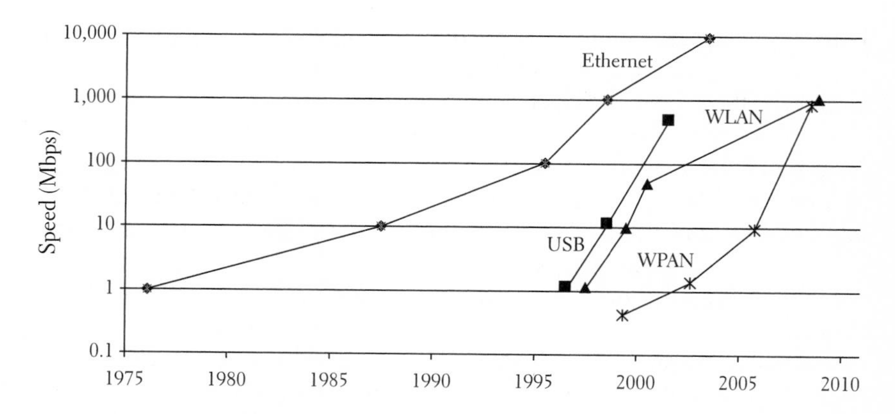

FIGURE 2.6 Improvements in data rates

SOURCE: Reprinted with permission of the International Solid-State Circuits Conference from the 2010 proceedings.

used to deliver data at a similar level of speed as long as ICs are available that can modulate a signal at this frequency. However, these ICs probably won't be available for at least decades if not longer.

CONCLUSIONS

The concept of a technology paradigm is a powerful way to analyze the potential of new technologies. It provides far more insight into this potential than do learning or experience curves. By merely providing short statements concerning a technology's concepts, trajectories, and basic methods of improvement, one can better understand its potential for improvements, the possible limits to this potential, and the technology's advantages and disadvantages with respect to other technologies.

Several conclusions emerge from this chapter. First, technological trajectories define the directions of advance for a technology, and there are four broad methods of achieving advance: (1) improving the efficiency with which basic concepts and their underlying physical phenomena are exploited; (2) radical new processes; (3) geometrical scaling; and (4) improving "key" components. This chapter focused on methods 1 and 4. The first method is relevant for engines, electricity generation, lighting and displays, and telecommunications; it includes finding materials that better exploit basic concepts and their underlying physical phenomena.

Finding better materials is particularly important for the low levels of a "nested hierarchy of electronic subsystems," such as batteries, lighting, displays, vacuum tubes, ICs, and memory storage. The realization and exploitation of each physical phenomenon that forms the basis of these technologies required a specific type of material, and finding the best material took many years. The best material exploited the physical phenomenon more efficiently than other materials did, and this higher efficiency often led to lower costs because fewer materials were needed. Strong bases of scientific knowledge facilitated the search for the best materials; without broad and deep knowledge bases, this search would have taken much longer.

On the other hand, improvements in technologies higher in an overall nested hierarchy of subsystems primarily depended on improvements in "key components." For example, improvements in transportation equipment depended on improvements in engines, and improvements in electronic products such as radios, televisions, computers, and mobile phones depended on improvements in electronic components such as vacuum tubes, transistors, and ICs. Furthermore, strong bases of scientific and technical knowledge facilitated the effective use of these components (this is addressed in more detail in Part II).

A second conclusion from this chapter is that dramatic improvements in performance and price rarely occurred following the emergence of technological discontinuities for most engines, turbines, energy storage devices, lighting, computing, and wireless systems. Although diesel and internal combustion engines might be defined as exceptions, it is probably more accurate to say that they had more potential for improvement than other engine types. Instead, evidence from this and other chapters (and other sources, such as analyses of energy transmission and wireless spectral efficiency[31]) suggests that improvements in performance and/or price occur in a rather smooth and incremental manner over multiple generations of discontinuities. These incremental improvements enable one to analyze new technologies and roughly understand their near-term trends in performance and/or price.

Consistent with other sources,[32] a third conclusion from this and subsequent chapters is that fully exploiting a technology paradigm requires improvements in a broad range of components. In the 18th and 19th centuries, it was improvements in the performance of basic metals, metal parts, and manufacturing processes that were needed before the full potential of technology paradigms for steam engines, locomotives, steamships, internal combustion engines, automobiles, electricity-generating equipment, and electric discharge tubes could be realized. In the 20th century, not just improvements in vacuum tubes, ICs, magnetic storage, and LCDs were needed before the technology paradigms for telephones, radio and television broadcasting, mobile phones, and the Internet could be fully exploited; also necessary were improvements in manufacturing processes for these vacuum tubes, ICs, magnetic disks, and LCDs. In the 21st century, further improvement in both these components and their processes—although these processes are now called nanotechnology—will be needed to exploit the full potential of new technology paradigms. As the processes used to make ICs and magnetic disks have reached submicron levels, *nanotechnology* has emerged that is in reality merely an extension of existing processes. Nevertheless, the ability to improve nanoprocessing, and thus the importance of understanding its technology paradigms, will likely have a large impact on the ability to create a wide variety of new industries.

A fourth conclusion is that it is possible to define much richer technology paradigms than are described in this chapter, which will help us understand potential opportunities for new and existing firms. This is partly the subject of Part II for several past technologies and Part IV for several existing and future technologies, where the focus is primarily on the relationship between improvements in components and discontinuities in systems. Focusing on the future, engineers, scientists, new business development personnel, policy analysts, and even students can use the technology paradigm to analyze directions of advance within a technology, and use this analysis to identify potential opportunities. Although a background in science and technology helps, curiosity and an open mind are probably more important prerequisites for such an analysis. By collecting data on cost and performance, and on their rates of change, for existing and new technologies,[33] one can analyze when a new technology might provide a superior value proposition and help determine market segments for it.

Finally, defining limits for a technology is inherently difficult. In part because these limits depend on both the performance and cost of supporting components and levels of scientific knowledge, it is easy to overlook potential solutions. For example, the end of Moore's Law has been predicted for almost twenty years and yet it still goes on. Many would argue that the only real limits are those of physics and our understanding of it, and even this understanding is often questioned. For example, the speed of light was being questioned as this book went to press. David Deutsch may be right that this is "the beginning of infinity." As long as humans search for better "explanations," there are no limits.[34]

3

Geometrical Scaling

Geometrical scaling is one type of technological trajectory within a technology paradigm, and some technologies benefit more from it than do others. Those that benefit over a broad range of either *larger* or *smaller* scale typically have more potential for improvements in cost and performance than do technologies that do not benefit at all or that benefit only over a narrow range. Technologies that benefit from increases in scale do so because output is roughly proportional to one dimension (e.g., length cubed, or volume) more than are costs (e.g., length squared, or area), thus causing output to rise faster than costs as scale is increased. For technologies that benefit from reductions in scale, they do so because reductions in scale lead to both increases in performance and reductions in cost and the benefits can be particularly significant. For example, placing more transistors or magnetic or optical storage regions in a certain area increases speed and functionality and reduces both power consumption and size of the final product. These are typically considered improvements in performance for most electronic products; they also lead to lower material, equipment, and transportation costs. The combination of increased performance and reduced costs as size is reduced has led to exponential changes in the performance-to-cost ratio of many electronic components.

This chapter uses data on cost versus scale, and the fact that the scales of some technologies were dramatically increased or decreased over time (see Tables 3.1 and 3.2), to show the potential benefits from geometrical scaling. Even without cost data, that the scale of a technology was dramatically increased or decreased over time suggests that there were large benefits from scaling. The chapter details these benefits for four types of technologies:

TABLE 3.1

Types of geometrical scaling

Technology	Subtechnology	Larger or smaller	Geometrical scaling
Production equipment	Continuous-flow plants	Larger	Cost is function of surface area of pipes/reaction vessels; output is function of volume
	Furnaces and smelters	Larger	Cost/heat loss are functions of surface area; output is function of volume
	Discrete-parts and assembly plants	Larger	Cost does not rise to extent machine speeds do
Engines/energy	Steam engines	Larger	Costs are function of surface area (e.g., cylinder, piston, boiler); output is function of volume. Larger sizes enable higher temperatures, leading to higher efficiencies
	Steam turbines	Larger	
	ICE	Larger	
Transportation equipment	Ships	Larger	Cost per passenger/freight-mile is function of sphere's (rather elongated) surface area; number of passengers/freight is function of sphere's volume
	Vehicles (e.g., buses, trucks)	Larger	
	Passenger aircraft	Larger	
Electronics	ICs	Primarily smaller	Smaller feature sizes reduce costs per transistor/memory cell and increase performance (density/speed)
	Magnetic and optical storage	Primarily smaller	
	LCDs	Smaller (thinness) and larger (equipment)	Costs fall as thinner films/larger areas processed in larger/higher-volume equipment
Clean energy	Solar cells		
	Wind turbines	Larger	Output rises with blade diameter squared; costs rise with diameter

SOURCE: Haldi and Whitcomb (1967); Levin (1977); Sahal (1985); Hirsh (1989); Freeman and Louçã (2002); Lipsey, Carlaw, and Bekar (2005); and Winter (2008).

(1) production equipment; (2) energy systems; (3) transportation equipment; and (4) electronics. For some of these technologies, it argues that when a component has a large impact on the cost and performance of a system and when it receives benefits from geometrical scaling over a broad range, those benefits can drive rapid improvements in the cost and performance of the system.

Most important, in combination with Chapter 2, this chapter shows that geometrical scaling can better explain the drivers of cost reduction than can the conventional wisdom concerning cumulative production, including why some technologies experience greater improvements in cost and performance than do others. Managers, engineers, professors, and even students can use the notion of geometrical scaling to better analyze the extent to which

TABLE 3.2

Extent of geometrical scaling (approximate figures)

Technology	Subtechnology	Examples of early scale	Examples of later scale
Production equipment	Continuous-flow plants	Less than 10,000 tons of ethylene/year in 1945	500,000 tons of ethylene/year by 1968
	Blast furnace	Single ton/day in 1700 for steel	10,000 tons/day by 1990 for steel/copper
	Discrete-parts and assembly plants	Hundreds of cars/year in 1900; 18,000 in 1909	1,000,000 cars/year at Highland Park in 1920
Engines/energy	Steam engines and steam turbines	10 HP in 1817; 10 MW by 1905	1,300 MW (1.3 million HP) by 1960s
	ICEs	¾ HP in car engine in 1885	90,000 HP in marine engine in 2009
	Jet engines	1,250-pound thrust in 1942	127,000-pound thrust in 2009
Transportation equipment	Passenger ships	10 mph and 48 passengers in 1838	26 mph and 2,150 passengers in 1984
	Oil tankers	1,807 tons in 1878	564,000 tons in 1979
	Passenger vehicles	Single passenger in 1885	300 passengers in 2009
	Passenger aircraft	12 passengers, 190 mph in 1931 (DC-1)	900 passengers, 560 mph in 2005 (A380)
Electronics	ICs	1 transistor in 1958	2 billion transistors in 2008
	Magnetic disks	2,000 bits/square inch in 1956	150 million bits/square inch in 2007
	LCDs	0.08 square meter in early 1990s	5.7-meter diagonal (16-square meter) panels in 2009
Clean energy	Solar cells	Not available	5.7-meter panels in 2009
	Wind turbines	15-meter blade diameter in 1985	124-meter blade diameter in 2007

SOURCE: Price per area of semiconductors was calculated using data on die size from ICKnowledge (2009), and price per transistor from Kurzweil (2005). Price per area of LCDs is from Gay (2008). Ratios of new to old prices per area are calculated using changes in efficiency and price reported in Nemet (2006). Interpolations and extrapolations are done for some years.

NOTE: HP, horsepower; MW, megawatt; mph, miles per hour; IC, integrated circuit; LCD, light-emitting diode.

improvements might occur in the future and thus when a new technology might provide a superior "value proposition" to some set of users.

GEOMETRICAL SCALING IN PRODUCTION EQUIPMENT

Geometrical scaling in production equipment is closely related to economies of scale, and some might argue that they are the same phenomenon. This book distinguishes between the two because geometrical scaling can help us understand the extent to which new types of processes might benefit from economies of scale. With economies of scale, unit costs drop as the equipment's fixed and operating costs are amortized over a larger volume—at least until the capacity of the equipment is reached and additional units of the same equipment must be installed. When production equipment benefits from geometrical scaling, larger-scale equipment has lower unit cost than smaller-capacity equipment on a per-output basis. Theoretical and empirical analysis suggests that the benefits from increasing the scale of production equipment are greater in chemical than in discrete-parts manufacturing where furnaces and smelters occupy an intermediate position. It is through these greater benefits that continuous-flow factories such as those that produce chemicals, paints, and drugs benefit more from economies of scale than other types of manufacturing.[1]

Furthermore, radical new processes are more common in chemical plants, smelters, furnaces, and even integrated-circuit (IC) production than in discrete-parts production. For example, the Bessemer process, the basic oxygen furnace, and continuous casting for steel; the Haber-Bosch process for ammonia; the float glass process for glass; the Hall-Héroult process for aluminum; and many other processes have had large impacts on the cost of chemicals and basic materials and have been extensively documented.[2] The introduction of these radical processes, which are very rare, was always followed by increases in their scale.

The production of organic and inorganic chemicals, plastics, paints, and pharmaceuticals (sometimes called continuous-flow processes) largely consists of pipes and reaction vessels where the radius of the pipes and reaction vessels is their key dimension. Many scholars have noted that the costs of pipes vary as a function of radius (which determines the surface area of a pipe) whereas the output from a pipe varies as a function of radius squared (the square of the radius impacts on the volume of flow). Similarly, the costs of reaction vessels vary as a function of radius squared (which determines the surface area of a sphere) whereas their output varies as a function of radius cubed (which determines the volume of the sphere). This means that output rises faster than does cost as radius is increases and empirical analyses have confirmed these advantages where the capital costs of continuous-flow factories are a function of plant size to the *n*th power, where *n* is typically between 0.6 and 0.7.[3] For example, the capacities of ethylene and ammonia plants were increased by about 50 and 10 times, respectively, between the early 1940s and 1968. For ethylene plants, this meant that capital costs on a per-unit basis in

1968 had fallen to about 25 percent of their early 1940s level.[4] In other analyses, it has been found that the cost of catalytic cracking (for gasoline) dropped by more than 50 percent for materials, 80 percent for capital and energy, and 98 percent for labor between an installation in 1939 and one in 1960.[5]

Similar arguments can be made for furnaces and smelters, which occupy an intermediate position between continuous-flow and discrete-parts production in terms of the benefits from geometrical scaling. Furnaces and smelters are used to process metals and ceramics, and benefits from a larger scale exist in their construction and operation. This is because, as with the previous examples, the cost of constructing a cylindrical blast furnace is largely a function of surface area while the output is a function of volume.[6] Construction costs include both material and processing, where, for example, the cost of welding together a heat furnace is proportional to the length of the seams while capacity is a function of the furnace volume. Similarly, the heat loss from a blast furnace or smelter is proportional to the area of its surface while the amount of metal that can be smelted is proportional to the cube of the surface sides.[7]

Although the cost of virtually every processed metal has dropped over the last 100 to 150 years, only two examples are provided here. First, the construction cost per ton of steel capacity dropped by eight times as the capacity of a basic oxygen furnace was increased by about 80 percent in the 1950s and 1960s, with the largest plants producing more than 2 million tons per year.[8] Second, the energy per kilogram of finished aluminum produced using the Hall-Héroult process fell by about 75 percent between 1890 and 2000 as the scale of aluminum plants was increased by about 300 times. In this case, the size of the "cells" used to produce aluminum are typically measured in terms of electrical current (i.e., amps), and increases in the current for a single cell led to lower energy use and lower costs. By the year 2000, the energy use per kilogram was only about 50 percent higher than its theoretical minimum—in other words, about one-third of the energy came from heat loss.[9]

The potential benefits from increasing the scale of production equipment in discrete-parts production are probably much smaller than they are from increasing the scale of furnaces, smelters, or continuous-flow manufacturing plants. Although the cutting speeds of lathes and boring machines were increased so that these machines could produce more parts per time and equipment cost than smaller and slower machines,[10] it is harder to increase the scale of these processes, as well as the scale of those used to form or assemble small parts, than it is to increase the scale of pipes and reaction vessels, partly because individual parts must be moved and processed.[11] It is even harder to increase the scale of or even automate the processes used to assemble shoes or stitch together apparel than it is to increase the scale of metal cutting, forming, and assembling. Thus, mechanical products such as automobiles, appliances, and bicycles benefit more from increases in scale than do shoes and apparel.

In summary, the chapter's characterization of cost reductions in manufacturing plants has a different focus from that of most characterizations. For example, in Utterback's characterization, innovations in products give way to innovations in processes as volumes increase, a so-called "dominant design" emerges,[12] and firms reorganize their factories and introduce flow lines and special-purpose manufacturing equipment to benefit from increases in volume. Some students of technology change interpret Christensen's model of disruptive technologies in this way: the expansion of production that occurs once a niche is found enables costs to drop and allows low-end products to displace mainstream products.[13] However, the fact that, one, some discrete-parts products (e.g., shoes and apparel) benefit less from larger scale and have experienced fewer reductions in cost than have other discrete-parts products and that, two, discrete-parts products benefit less from larger scale and have experienced fewer reductions in cost than have continuous-flow products (and those from furnaces and smelters) suggests that geometrical scaling is a better explanation than these characterizations.[14]

A second reason for focusing on geometrical scaling is that it enables us to better understand the improvements in supporting components that have enabled reductions in production cost. For continuous-flow plants, improvements in materials such as steel were needed so that large increases in the thickness of pipes and reaction vessels would not be needed to compensate for increases in pipe and vessel radius. As for furnaces and smelters, their scale was limited by a need to deliver a smooth flow of air to all of the molten metal where hand- and animal-driven bellows could only deliver a limited flow of air. Water-driven and steam-driven bellows allowed air to be injected with more force, so larger furnaces could be built.[15]

For all forms of manufacturing, improvements in tolerances and the availability of low-cost electricity were needed to implement larger-scale production equipment. In the late 19th century, improvements in machine tools such as lathes and boring machines led to tighter tolerances and better parts, and these parts were used to make better evaporators, mixers, blenders, and other machine tools.[16] Furthermore, a broader and deeper knowledge of machining led to new materials for machine tool bits so that tolerances could be improved and cutting speeds increased.[17] As for electricity, it became possible to use an electric current to drive a variety of chemical reactions in, for example, electrometallurgy, electrochemistry, and electrolysis.[18] Low-cost electricity also enabled machine tools to be driven by electric motors, which allowed new forms of factory organization in which the placement of machines no longer needed to be dictated by the system of shafts and belts that drove them.[19] Although electric motors did not benefit from increases in scale, their production and, more important, the generation of electricity did benefit (see the next section). This caused the price of electricity to drop by more than 95 percent in inflation-adjusted terms in the United States between the late 19th century and 1973.[20]

GEOMETRICAL SCALING IN CONVENTIONAL ENERGY SYSTEMS

Benefits from increasing the scale of energy systems exist in steam, internal combustion, and jet engines, the largest steam engines are now thousands of times larger than they were 300 years ago (see Figure 3.1). Steam and internal combustion engines (ICEs) benefit from increases in scale because the output from a cylinder and a piston is roughly a function of volume (i.e., diameter squared), while costs are roughly a function of the external surface areas of the piston and cylinder (i.e., diameter); see Figure 3.2. Thus, as the diameter of the cylinder (and piston) is increased, the output of the engine increases as a function of diameter squared; costs rise only as a function of diameter. Steam engines also benefit from increases in the scale of the boiler: as the diameter of the boiler is increased, the output of the engine increases as a function of diameter cubed, while costs rise only as a function of diameter squared. These increases also facilitated higher temperatures and pressures, which contributed toward improvements in thermal efficiency (see Figure 2.1). Nevertheless, the benefits from increases in scale depended on advances in the science of thermodynamics and combustion[21] and on improvements in materials. Like pipes and reaction vessels, without improvements in steel and other materials, greater thicknesses of cylinder, piston, and boiler walls would probably have been necessary and may have eliminated the advantages of increasing the scale of engines.[22]

For example, consider the price per horsepower (HP) of modern ICEs and that for steam engines in 1800. In 2010,[23] Honda's 225-HP marine engine was 26 percent the price of its 2.3-HP engine. For steam engines in 1800, the price per HP of a 20-HP engine was one-third that of a 2-HP engine. The advantages of scale become even more apparent when one considers the size of the first internal combustion engine (only three-quarter HP), the largest one today (90,000 HP), and the largest steam turbines (1.3 million HP), which are the modern equivalent of steam engines. Extrapolating to these extremes suggests that the largest internal combustion engines would be less than 1 percent the price per HP of the smallest ones in current dollars.[24]

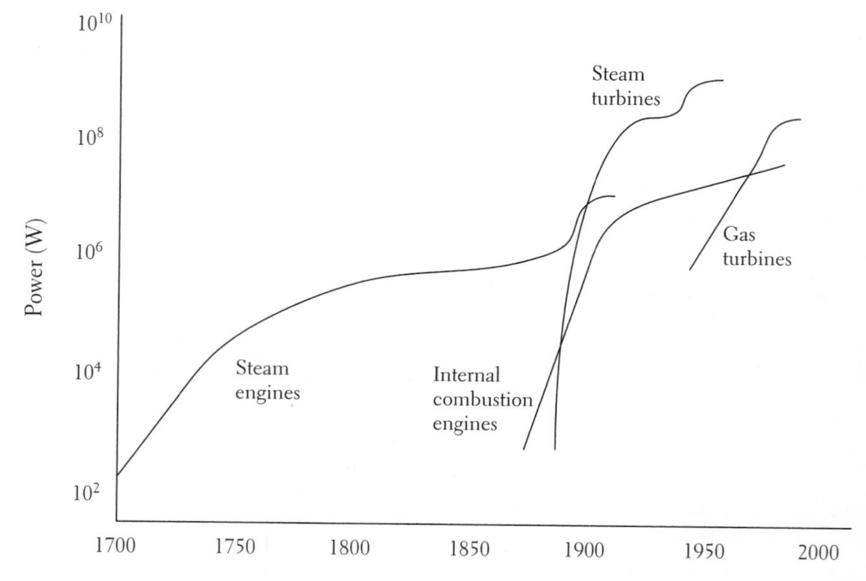

FIGURE 3.1 Maximum scale of engines and turbines

SOURCE: Republished with permission of ABC-CLIO, Inc., from *Energy Transitions: History, Requirements, Prospects*, Vaclav Smil, © 2010; permission conveyed through Copyright Clearance Center, Inc.

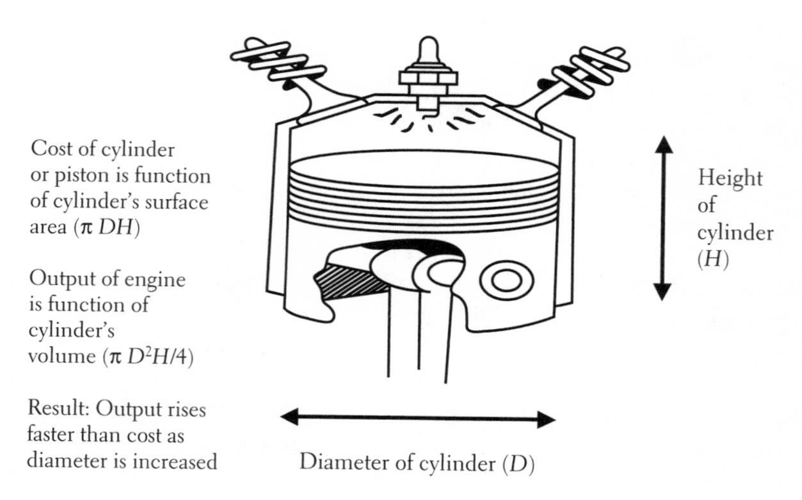

FIGURE 3.2 Examples of geometrical scaling with engines

SOURCE: http://www.wpclipart.com/transportation/car/parts/piston_cylinder.png.html.

The falling price of electricity is also largely attributed to increases in the scale of steam turbines, boilers, generators, and transmission lines, leading to reductions in the capital cost per unit of output. For example, the cost per installed capacity dropped from about $78 per kilowatt for a 100-megawatt coal-fired plant to about $32 per kilowatt for a 600-megawatt plant, both in 1929 dollars. Nuclear plants experienced similar cost reductions as their scale was increased.[25] The falling capital costs per output, which were probably even larger for increases in scale from single-digit megawatt plants to 100-megawatt plants, caused the price of electricity to fall from $4.50 per kilowatt-hour in 1892 to about $0.09 by 1969 in constant dollars.[26] From 1969, however, fewer increases in scale were implemented, and many observers now argue that the United States had already implemented too much scale, and more scale than in Europe because of institutional differences. The cost of electricity in the United States has risen since 1969, where excessive scale is cited to a similar extent, as is increased fuel cost. Some argue that smaller generating plants such as combined-cycle gas turbines that produce both heat and electricity have much higher efficiencies and thus lower costs than larger coal-fired power plants.[27]

A third example of geometrical scaling can be found in jet engines, which have benefited from increases in scale for some of the same reasons that steam turbines have. Like the pipes in a chemical plant, a jet engine's combustion chamber benefits from increases in scale in that costs rise with surface area while output rises with volume. Also, larger engines enable higher temperatures and higher temperatures enable higher thermal efficiencies, with jet engines operating at much higher temperatures than other engine types.[28] Improvements in materials and dramatic increases in complexity were necessary to implement these larger engines that operate at higher temperatures.[29]

GEOMETRICAL SCALING IN TRANSPORTATION EQUIPMENT

Benefits from increases in the scale of transportation equipment exist for many of the same reasons as those for engines, furnaces, and continuous-flow manufacturing plants. The carrying capacity of ships (including tankers), buses, trucks, and to some extent aircraft rises with volume (i.e., cube of a dimension) while cost rises with surface area (i.e., square of a dimension).[30] The surface area could be that of a ship's hull, a bus' body, a truck's rear bed, or an airplane's fuselage, where each is similar to the shape of a rather elongated sphere, although aircraft have wings and truck beds are more like cubes than spheres. Furthermore, the speed of transportation equipment is also a function of size, partly because the engines benefit from increases in scale.

For example, Table 3.3 compares the price per capacity of small- and large-scale oil tankers, freight vessels, and aircraft. The prices of the largest oil tankers and the largest freight vessels are 59 percent and 50 percent lower than the smallest ones on a per-ton basis, respectively. The prices of the largest commercial aircraft are 14 percent lower than those of the smallest on a per-capacity (passenger) basis. Not only are labor and fuel costs also less on a per-capacity basis for the larger oil tankers, freight vessels, and aircraft than those for the smaller ones; the advantages of scale become even more apparent when one considers that the first oil tankers could only handle 1807 tons (in the late 19th century) and the first commercial aircraft, the DC-1 (in the early 1930s), could only carry 12 passengers. Extrapolating suggests that today's largest oil tanker is almost one-twentieth the price per ton of an 1807-ton tanker and that the A380 has a price per passenger almost one-half that of the DC-1.[31] These are some of the reasons that the price per ton-mile of rail freight in the U.S. dropped by 88 percent between 1890 and 2000, that the price of airline tickets has dropped dramatically over the last few decades, that the share of U.S. GDP for transport dropped by more than 50 percent between 1860 and 2000 (and by more than 75 percent if airline travel is not considered),[32] and that both global trade and global travel grew faster than did overall economic output in the 20th century.

Like the previous examples, these increases in scale did not come easy. Advances in thermodynamics, combustion, and fluid flow, and improvements in materials (i.e., components), were necessary for the benefits from larger scale to emerge. Improvements in steel and steam engines in the 19th century were needed to make larger ships, and increases in the size of ports and canals are

TABLE 3.3

Current prices per capacity for large- and small-scale oil tankers, freight vehicles, and aircraft

Scale	Dimension	Oil tankers	Freight vessels	Aircraft
Large	Price	$120 million	$59 million	$346.3 million (A380)
	Capacity	265,000 tons	170,000 tons	853 passengers
	Price per capacity	$453 per ton	$347 per ton	$405,979 per passenger
Small	Price	$43 million	$28 million	$62.5 million (A318)
	Capacity	38,500 tons	40,000 tons	132 passengers
	Price per capacity	$1116 per ton	$700 per ton	$473,348 per passenger

SOURCE: See UNCTAD (2006); also see https://docs.google.com/viewer?a=v&q=cache:hbqTkQ9iPRwJ:www.airbus.com/presscentre/corporate-information/key-documents/?eID%3Ddam_frontend_push%26docID%3D14849+airbus+aircraft+2010+list+prices+a318&hl=en&gl=sg&pid=bl&srcid=ADGEESg6wOxcKDNEpSDxx1ipIMm-tL-ZwByNA2inVWLwLtj1SHNpn9BJFT0Qteowo8cWvdXU2RjSnHjQst54JNILnoiN-PzFIM9hW30fmGfJ9guFv9rz86Jz14QBAwbxV-p-zR9oN9m2S&sig=AHIEtbQ-aPc1T2JXZW74p-1g8sMYKk738yA.

still being implemented. Larger buses required improved steel; better ICEs required improvements in materials such as aluminum and plastics. Larger aircraft required improvements in aluminum, jet engines, and more recently composites—aircraft weight has dropped significantly over the last 20 years as the strength-to-weight ratio of materials has increased several times.[33] On the other hand, limits to larger scale are appearing in buses and aircraft, and they are probably right around the corner for sh

A SPECIAL CASE OF GEOMETRICAL SCALING IN ELECTRONIC COMPONENTS AND SYSTEMS

Electronic components might be considered a special case of geometrical scaling, for several reasons. First, improvement in the performance and cost of ICs and magnetic and optical storage systems are sometimes called "exponential" since there is a short time period for each doubling. This frequent doubling results in multiple "orders of magnitude" in improvements over longer time periods (see Table 3.4) and thus is much higher than anything seen in human history.[34] Second, some of these improvements have had a large impact on the performance of electronic systems such as computers, mobile phones, and the Internet. Improvements in these and related components were the primary driver for the emergence of personal and portable computers, mobile phones, and Internet industries, as well as new industries within the broadcasting, telecommunication, health care (including biotechnology), education, and financial sectors. (These are addressed in more detail in subsequent chapters.) For example, improvements in computers, robotics, and the Internet, which depended on improvements in electronic components, have enabled high-throughput screening of drugs, systems modeling of biological processes, computer-aided tomography (CAT) and other forms of medical imaging, online universities, and automated automatic algorithmic trading by, for example, hedge funds of stocks, bonds, currencies, and derivatives (e.g., credit fault swaps).[35]

Third, geometrical scaling in ICs and magnetic and optical discs primarily involves smaller scale. While previous examples involve increasing scale, exponential improvements in the cost and performance of electronic components come primarily from dramatically reducing scale, a phenomenon first noted by Nobel laureate Richard Feynman in a 1959 speech entitled "There's Plenty of Room at the Bottom: An Invitation to Enter a New Field of Physics."[36] Feynman could not have been more insightful about the future. However, the issues he addressed, and the benefits that have emerged from smaller scale in many electronic components, have implications for technologies that

TABLE 3.4

Geometrical scaling and rates of improvement for selected components

Component	Measure of performance	Rates of improvement
ICs (and some bioelectronic ICs and MEMS)	Feature size	>2 OoM (500 times) in 40 years
	Defect density	>3 OoM in 40 years
	Die size	>30 times in 25 years
	Number of transistors per chip	9 OoM in 50 years
Semiconductor/LED/LCD-manufacturing equipment	Cost per area of ICs	See Table 3.5
	Cost per area of LCDs	See Table 3.5
Hard disk platters	Areal storage density	5 OoM in 40 years
Magnetic tape	Areal storage density	5 OoM in 45 years
Optical discs	Capacity	10 times in 10 years
	Transfer rates	3 OoM in 10 years
Glass fiber	Capacity (bits/second)	5 OoM in 20 years
	Cost per bit	6 times reduction in 25 years

SOURCE: Kurzweil (2005); Molstad, Langlois, and Johnson (2002); and author's analysis.

NOTE: IC, integrated circuit; MEMS, microelectronic mechanical systems; OoM, orders of magnitude; LED, light-emitting diode; LCD, liquid crystal display.

benefit from both larger and smaller scale such as LCDs and solar cells. The next section summarizes the key aspects of miniaturization in electronic components and their impact on electronic systems.

Benefits from Smaller Scale in Electronic Components

As noted in Chapter 2, the technology paradigm for ICs involves reducing the distances between the inputs and output of a transistor, the width of metal lines connecting transistors, and the thicknesses of various layers. Smaller line widths and smaller material thicknesses increase an IC's speed, density, and functionality (e.g., number of transistors per chip), and they reduce size and costs. Material and energy costs are roughly a function of surface area and thickness, and so reductions in feature size lead to lower costs as long as improvements in equipment occur. These improvements partly depend on broadening and deepening the knowledge of solid-state physics and related scientific disciplines.

Equipment for defining and etching patterns and depositing, diffusing, and implanting materials was borrowed from a diverse set of industries in the 1950s, including printing, aerospace, and nuclear energy, before the semiconductor industry became the main driver of these improvements in the same decade. Improvements in the form of photolithographic equipment, plasma etchers, and ion implantation equipment reduced defect densities and feature sizes, with the reduction in defect densities enabling a 30-fold increase in die size over the last 25 years. Reduced feature sizes and to a lesser extent larger die sizes have increased the number of transistors that can be placed on a single IC chip, according to what is often called Moore's Law.[37] Similar arguments about the benefits of reduced feature sizes are made for microelectronic mechanical systems (MEMSs) and bioelectronic ICs in Chapter 6.

Part of Moore's Law is also driven by the benefits from increasing the scale of semiconductor-manufacturing equipment itself. First, just like the chemical processes and furnaces discussed previously, output has risen faster than equipment costs as equipment size has grown. The reason is that processing time (the inverse of output) has fallen as the volume of gases, liquids, and reaction chambers has increased, while costs have risen as a function of the equipment's relevant surface area. Second, the ability to process multiple ICs on a single wafer, whose size has been increased many times in the last 50 years, supports the first reason because loading time and the costs of loading equipment do not increase much with increases in wafer size and because there are smaller "edge effects" with larger wafers.[38] Third, the techniques for reducing the line width of patterns on IC chips have required manufacturers to reduce the thickness of the deposited (and later patterned) materials, which reduces the cost of materials and processing time. The result is that costs per transistor (or LED or memory cell), capital costs per transistor, and even to some extent *costs per area of a silicon wafer* (see Table 3.5) have fallen over the last 50 years even as the cost of fabrication facilities has increased.[39]

On the other hand, the rising cost of fabrication facilities cannot go on forever,[40] and many observers see diminishing returns to smaller scale in ICs and

TABLE 3.5

Cost reductions for ICs, LCDs, and solar cells

Technology	Dimension	Time frame	Ratio of new to old cost
Transistors/ICs	Price/transistor	1970–2005	1/15,000,000
	Price/area	1970–2005	1/20
	Price/area	1995–2005	1/5.7
LCDs	Price/area	1995–2005	1/20
Solar cells	Price/watt	1957–2003	1/500
	Price/watt	1975–2001	1/45.4
	Price/area	1970–2001	1/37.0
	Price/area	1995–2001	1/3.42

SOURCE: Price per area of semiconductors was calculated using data on die size from IC-Knowledge (2009), and price per transistor from Kurzweil (2005). Price per area of LCDs is from Gay (2008). Ratios of new to old prices per area are calculated using changes in efficiency and price reported in Nemet (2006). Interpolations and extrapolations are done for some years.

NOTE: IC, integrated circuit; LCD, liquid crystal display.

argue the need for a new technology paradigm for them. For example, if one were to plot increases in the number of transistors per chip versus R&D effort at the industry level, as proposed more than 25 years ago by Richard Foster[41] and not versus time, as is almost always the case — one would see diminishing returns from R&D. Not only has the rate of improvement in the number of transistors per chip slowed in the last ten years, but the industry's R&D effort has increased, with the rising cost of equipment just one example. The need for a new technology paradigm can be seen in firms searching for new concepts in photolithographic equipment, interconnect, and "information processing and storage devices" themselves, as was mentioned in Chapter 2.[42]

Similar stories can be told about magnetic and optical storage. As with ICs, reducing the size of the storage cells for magnetic cores, drums, disks, tape, or optical discs increases density and speed and lowers costs, as long as the relevant advances in science are achieved and improvements in equipment are made.[43] Improvements in sputtering equipment and scientific advances in, for example, giant magneto resistance (published in late 1980s) enabled smaller and smaller storage cells on magnetic disks. For optical discs, reductions in the wavelength of light emitted by semiconductor lasers are needed to reduce storage cell size and the space used by light in a fiber-optic cable. As mentioned in the last chapter, because reducing the wavelength is becoming more difficult, optical discs will probably be replaced by other technologies.

Diminishing Returns to Larger Scale in Electronic Systems

The benefits from smaller scale in ICs and magnetic and optical discs have had a dramatic impact both on the performance of electronic systems and on the emergence of discontinuities in them, which is the focus of Part II. One reason for this is that electronic systems probably benefit much less from increases in scale than do other technologies such as production and transportation equipment.[44] For example, consider computers. In the 1940s, Grosch noted that the cost of computing power only increased as the square root of processing power, and thus the cost per instruction per second declined as the scale of computers was increased. Analyses by computer scientists in the 1960s and 1970s confirmed this,[45] which suggests there are some benefits from increasing scale, perhaps because the cost of support circuitry (and input-output devices) drops as computer size grows.

However, as new types of computers such as minicomputers and PCs began to diffuse, new analyses found that smaller computers had a lower computing cost than larger ones did. The reason is that smaller computers typically use more standard components such as high-volume ICs and thus have lower development costs.[46] For example, while the largest mainframes used custom ICs in order to maximize processing speeds, smaller mainframes and minicomputers used off-the shelf ICs; PCs took this one step further with standard microprocessors. That standard high-volume ICs had lower costs than did custom ICs is partly because larger-scale equipment could be used to produce them, but more because development costs on a per-unit basis were much lower for standard microprocessors and PCs than for custom ICs and large mainframes. A lack of benefits from increasing scale, lower development costs per unit volume of smaller computers, and other factors meant that the success of minicomputers and PCs was largely inevitable. It was only a matter of when these new computers could provide sufficient levels of performance at a low enough price (see Part II).

Diminishing returns to scale also exist in software. Although many focus on the low cost of duplicating software, which comes from the falling cost of memory and microprocessor ICs, the cost of both developing and implementing software in corporate settings has increased faster than has program size.[47] These diminishing returns are one reason that modular design is so widely used in electronic systems and that the levels of modular design continue to increase. For example, the emergence of open-source software, software as a service (SaaS), and utility/cloud computing represents growing levels of modular design that are partly driven by the diseconomies of scale in software systems. Furthermore, this is one reason that Nicholas Carr[48] and others expect SaaS and utility/cloud computing to replace conventional software and computing. As discussed in Chapter 9, SaaS and utility/cloud computing have much lower implementation costs than does conventional software, and improvements in Internet speed, which are primarily from the geometrical scaling of electronic components, are reducing their disadvantages as compared to conventional software.

Benefits from Both Larger and Smaller Scale in LCDs

Like ICs, LCDs benefit from increases in the scale of their substrates and production equipment. Firms have increased the size of substrates, which are cut up into LCDs for final products such as televisions, because larger substrates and their production equipment have lower equipment costs per output. Larger substrates are analogous to larger IC wafers, and their technologies and production equipment are also similar partly because so-called active-matrix LCDs use transistors to control the output of individual pixels.[49] However, in addition to differences in materials, the primary difference between the technology paradigms for ICs and LCDs is that smaller feature sizes (except for material thickness) are far less important and increases in screen or "panel" or substrate size (wafer size for ICs) are far more important for LCDs.[50] Large LCD production equipment (which processes large substrates) has lower equipment costs per output than small equipment does because it can more quickly handle and process substrates and produces smaller "edge" effects.[51]

Analyses of costs support this argument. First, as shown in Table 3.5, reductions in the cost per area of LCDs (as with solar cells) through 2005 were larger than those for ICs in part because suppliers of LCDs placed more emphasis on cost per area than did suppliers of semiconductors (the latter emphasized cost per transistor). Second, the output (substrate area per hour) per dollar of capital costs for one type of LCD manufacturing equipment increased by 8.5 times as the substrate size increased by almost 16 times, from 0.17 (Generation II) to 2.7 square meters (Generation VI),[52] which is roughly consistent with the data in Table 3.5.

Third, the capital cost per area, this time for a complete facility, dropped by 36 percent as substrate size increased from 1.4 (in Generation V) to 5.3 square meters in Generation VIII.[53] Generation XI panels are now 10.5 square meters. Fourth, the most important material in LCDs, glass, also benefits from increases in the scale of substrate and production equipment.[54] Finally, the benefits from larger substrate size can be seen in the fact that the production costs for LCD televisions were 40 percent the levels in 2011 that they were in 2001 on a per meter squared basis.[55] All of this data suggests that LCD substrate size and production equipment greatly benefit from increases in scale and that the costs of solar cells also drop as their substrate size increases.

CONCLUSIONS

This chapter showed how geometrical scaling can drive large improvements in cost and performance. While many argue that these improvements come from increases in cumulative production volumes and changes from product to process innovations as dominant designs emerge,[56] these theories do not explain why some technologies experience greater improvements than others. In combination with the other three methods of achieving improvements, which are discussed in Chapter 2, this chapter demonstrated that geometrical scaling provides a more fine-grained explanation for these differences between technologies than conventional wisdom can give us. For larger scales, benefits from geometrical scaling primarily exist when output is roughly proportional to one dimension (e.g., length cubed or volume) more than are costs (e.g., length squared or area), causing output to rise faster than costs as scale is increased. For smaller scale, benefits from geometrical scaling primarily exist when performance (speed and functionality) rises and costs fall as reductions in size are achieved. For both smaller and larger scale, realizing these benefits requires changes in the product design and thus the above-mentioned change from product to process innovation does not occur in the technologies covered in this chapter.[57]

Second, this chapter's analysis showed how geometrical scaling in a single component can lead to dramatic improvements in the performance and cost of a system. For example, benefits from geometrical scaling in ICs have had a dramatic impact on the performance and cost of many electronic systems, including computers, mobile phones, and the Internet, even though many of these electronic systems themselves do not benefit from scaling. Benefits from geometrical scaling in engines have had a large impact on transportation equipment such as vehicles (e.g., buses and trucks), ships, and aircraft. In some cases geometrical scaling exists in both a "component" and a "system" and reinforces both. Examples include engines and transportation equipment and semiconductor-manufacturing equipment and ICs.

Third, there are probably trade-offs between the benefits from increased scale for a "system" and increased scale for the "system's" production equipment. For example, one could use a large engine or multiple small engines for a specific task. The large engine might benefit from its larger scale while the smaller engine might benefit from the use of larger production equipment that is made possible by the larger volumes associated with smaller engines.[58] The former case further highlights the problems associated with focusing on cumulative production as the primary driver of cost reductions. If cumulative production were the primary driver, firms would opt for the smaller engines in order to increase cumulative production.

Fourth, the role of geometrical scaling in discontinuities has implications for Christensen's theory of disruptive technologies. The belief that once a low-end innovation finds a niche it will naturally displace the dominant technology is reinforced by the conventional wisdom about cumulative production driving cost reductions.[59] However, the phenomenon is far more complex than this. One issue is whether the dominant technology exhibits "technology overshoot," and this depends on the market and the characteristics of users (see Part II). However, other common factors among Christensen's disruptive technologies are large improvements in cost and performance and geometrical scaling. Every disruptive technology described by Christensen in his 1997 book, including computers, hard disk drives, mechanical excavators, new information systems, and mini-mills, benefited from geometrical scaling either in the system itself (e.g., mini-mills) or in a key component in it.[60] Benefits from geometrical scaling exist in ICs for computers (and thus for information systems), magnetic platters for hard disk drives, and pumps and actuators (similar to geometrical scaling for engines) for mechanical excavators. Mini-mills are a little more complex than the other systems. Although they benefit from geometrical scaling for the same reasons that conventional steel mills do, they also use a different concept (the electric arc furnace) and benefit from the falling cost of electricity and the increasing availability of scrap metal.

Furthermore, most of the disruptive technologies covered in Christensen's other publications, including "The Great Disruption," *The Innovator's Prescription*, and *Disrupting Class*, also exhibit geometrical scaling. For example, all of the electronic technologies covered in "The Great Disruption" exhibit geometrical scaling in ICs, magnetic tape, or magnetic platters. For health care, advances in medical science and improvements in information technology (including imaging), of which the latter depends on scaling in ICs, permits health care to depend more on low-wage nurses (and patients) than on high-wage doctors. For education, improvements in information technology enable more customized learning via low-end methods such as eLearning or online universities than do existing methods.[61] The fact that most of Christensen's disruptive innovations involve large improvements in cost and performance primarily through geometrical scaling suggests that finding disruptive technologies is much more than just finding low-end innovations, which Christensen emphasizes in his latest book,[62] and others do as well.

Fifth, consistent with other sources (and the previous chapter), this chapter highlighted the challenges of accelerating the pace of scaling. Every instance of geometrical scaling described here was supported by and depended on improvements in components and advances in science. For example, deeper and broader scientific knowledge of chemistry, electricity, and materials for high-speed machining and improvements in special-purpose manufacturing equipment and electric motors supported increases in the scale of production equipment. Improved bellows for increasing the flow of air to furnaces facilitated increases in the scale of furnaces. Deeper and broader knowledge of thermodynamics and combustion, higher-tolerance parts, and better materials supported increases in the scale of engines. Advances in thermodynamics and fluid flow and improvements in steel, aluminum, plastics, and composites supported increases in the scale of transportation equipment.

Similar conclusions can be stated for reductions in scale. Deeper and broader knowledge of solid-state and plasma physics and improvements in many types of manufacturing equipment supported geometrical scaling in ICs, LCDs, and magnetic storage. Furthermore, much of this equipment supported both geometrical scaling at the smaller and larger scale; the latter occurred for many of the same reasons that equipment for continuous-flow industries benefit from increased scale. Geometrical scaling in optical discs and fiber optics also depended on deeper and broader knowledge of solid-state physics, but the relevant components were semiconductor lasers and sensors.

Finally, the ability to accurately explain reductions in cost and improvements in performance using the analyses here and in Chapter 2, rather than other methods, suggests that managers, engineers, professors, and students can use a technology paradigm, including the notion of geometrical scaling, to analyze the extent to which improvements might occur in the future and thus when a new technology, even a "disruptive" one, might provide a superior "value proposition" to some set of users or might diffuse to a broad set of users. Part II (as well as Part IV) analyzes these issues in more detail and focuses on systems in which components primarily benefit from geometrical scaling. An exception is Chapter 6's analysis of the semiconductor industry, where both components (equipment) and system (ICs) benefited from geometrical scaling but the system benefited much more.

II
When Do Technological Discontinuities Emerge?

Understanding when technological discontinuities and new industries might emerge is a critical issue for high-tech firms, governments, think tanks, and universities. Firms must decide when to introduce new products that can be defined as a technological discontinuity or as a new industry. Both firms and governments must decide when to fund research that may form the basis of a technological discontinuity or a new industry, but governments must also decide whether and when to introduce policies that might promote the introduction and diffusion of the products of a discontinuity. University professors and students must decide whether and when a new industry deserves time and study.

Although technological discontinuities do not always constitute new industries, Part II focuses on them, both because they often do form the basis of new industries and because there is wide agreement in the management and economics literature on actual discontinuities—when they were successfully commercialized and how they diffused. Technological discontinuities are typically defined and classified by the extent to which a new product, when compared to a previous one, involves changes either in the core concepts that form the basis of a product (or key component) or in the linkages among the product's key components.[1] Radical innovations change both the concepts and the linkages; architectural innovations change only the linkages; and modular innovations change only the core concepts of a single component. Although some scholars focus on a technology's impact on the linkages between a firm and the market,[2] these types of discontinuities, including so-called disruptive ones, can also be classified as either radical or architectural innovations.[3] To simplify the analysis, Part II focuses on the timing of radical and architectural innovations, partly because modular innovations are actually radical innovations at a lower level in a "nested hierarchy of subsystems.[4]"

We can define timing in terms of the time lag between the characterization of the concepts and/or architectures that form the basis of a discontinuity and the discontinuity's actual commercialization (and diffusion). This commercialization can only occur when a discontinuity, or any product for that matter, is adopted by some set of users because it provides a "superior value proposition" when compared to the previous technology. To achieve this, the discontinuity must have a lower price, higher performance, or some better combination of price and performance. Since users often evaluate products in terms of multiple dimensions of performance, and different users often emphasize different dimensions, analyzing why a discontinuity emerges and diffuses can be complex. For simplicity, we start with a single dimension of performance, a single user, and the emergence of a discontinuity for this user, and then consider multiple dimensions of performance, multiple users, and the diffusion of a discontinuity among these users.

THRESHOLDS OF PERFORMANCE AND PRICE

One way to analyze the emergence of a discontinuity and its adoption by a single user is in terms of minimum thresholds of performance and maximum thresholds of price. The performance of a new technology must exceed minimum thresholds of performance and/or fall below a maximum threshold of price before a user will consider purchasing it.[5] For example, many consumers will not even consider the purchase of smart phones until their price falls below a certain threshold. Similarly, many will not consider purchasing one until performance exceeds certain thresholds, such as specific amounts and types of content and applications and the ease of accessing them. Ease of access depends on the performance of the display, network and processing speeds, memory, overall user interface, and other factors.

In spite of multiple dimensions of performance for most technologies, specific dimensions often emerge as key. These include ease of access to content and applications for smart phones, processing speed for computers, storage capacity for magnetic recording equipment, recording density for magnetic platters and tape, the number of transistors per chip or feature size for semiconductors, and speed and capacity of optical fibers and wireless networks. Such specific dimensions are used for technologies because they represent an important part of a product's overall performance. They enable us to analyze the emergence of discontinuities in terms of when a single dimension of performance exceeds a minimum threshold and when price falls below a maximum threshold.

Part II focuses on systems for which single components determine these levels of performance and price more than do novel combinations of components. When a single component, and not a novel combination, is the bottleneck,[6] we can go beyond research that sees complementary technologies as the main source of a time lag between characterization of concepts/architectures and commercialization[7] and estimate the minimum thresholds of performance and maximum thresholds of price for the components that are needed before the system-based discontinuity reaches its performance/price thresholds. Furthermore, when a typically used measure of performance for a component also includes considerations of price or cost, we can focus just on a minimum threshold of performance for it; this is the case for most of the discontinuities addressed in Chapters 4, 5, and 6. For example, as described by Gordon Moore in his famous 1965 *Electronics* article,[8] which led to the term "Moore's Law," the number of transistors per chip includes considerations of both performance and price. Thus we can focus on the minimum threshold of performance that was needed in integrated circuits (ICs) before specific computers could be introduced. Similar observations can be made for the magnetic recording density of tape in Chapter 5 and semiconductor-manufacturing equipment in Chapter 6.

TRADE-OFFS THAT USERS AND DESIGNERS MAKE

Multiple dimensions of performance, multiple users, and the diffusion of a discontinuity across different users are now considered. We examine the trade-offs that designers and users make when they consider products, particularly new ones that can be defined as technological discontinuities. The idea of trade-offs builds from both Dosi's "technology paradigm" and theories of marketing. For example, hard disk drives, which store data in computers, have been closely analyzed by Christensen. From the perspective of a designer, there is a trade-off between disk capacity and disk diameter in that the only way to increase capacity at any given moment is to increase diameter. Over time, however, reductions in the size of magnetic storage regions led to improvements in the magnetic recording density of a platter (as discussed in Chapter 3), which made it possible to increase capacity without increasing diameter. Therefore, we can say that improvements in the magnetic recording density of the platter (i.e., a component) drove changes in the trade-offs that the designers of disk drives (i.e., the system) were making between capacity and diameter. These changes enabled them to introduce smaller-diameter disks with the same storage capacity as larger-diameter disks.

For users, one way to analyze trade-offs is through the notion of indifference curves and how they change over time. An indifference curve represents the set of points for two or more variables to which individual or multiple users (e.g., a market segment) are indifferent,[9] and thus it summarizes the trade-offs that a user or users are considering at a moment in time. While Christensen uses the term "technology overshoot" to analyze the diffusion of disruptive innovations, the notion of changes in trade-offs can be used instead. As products improve in performance and/or price, they may overshoot the needs of users and thus change the performance/price "trade-offs" that users make (as represented by an indifference curve). For example, as the recording densities of platters and thus the storage capacity of hard disks increased over time, capacity overshot the needs of some customers; we can therefore say that these capacity increases, which were driven by increases in recording density, changed users' trade-offs between capacity, price, and other factors. The result was that users began to place more emphasis on price and other dimensions of performance, and some high-end users switched from large disk drives with high storage capacities to smaller disk drives with lower storage capacities (and with lower prices and perhaps higher reliabilities).

One reason for characterizing this phenomenon as a change in trade-offs is that it does not restrict us to so-called low-end innovations, which Christensen's theory does. Many innovations begin as expensive high-end products that gradually become cheaper over time. Their introduction and falling price cause users to change the trade-offs between price and different dimensions of performance. Consider again the emergence of smart phones, but this time from the perspective of trade-offs. Before smart phones appeared, most consumers wanted small and light mobile phones. The introduction of inexpensive smart phones, which started with high-end business users, dramatically changed the trade-offs that most users made between size, weight, price, and other factors.

Chapters 4, 5, and 6 further investigate the impact of component improvements on the introduction and diffusion of several system-based discontinuities. They show that the rates of improvement in performance and price of components (and thus systems) were extremely high and thus probably had an impact equal to, if not larger than, the degree of heterogeneity or what some call the degree of preference overlap. If we plot the indifference curves for two market segments in a single graph, we can analyze the degree of preference overlap (or degree of heterogeneity) in an overall market.[10] A low preference overlap (high heterogeneity) leads to many different market segments, which facilitates the emergence of discontinuities but not their diffusion in the overall market. A high preference overlap leads to fewer segments; this homogeneity does not facilitate the emergence of discontinuities, but it does facilitate their diffusion once they capture a small part of the market.[11] However, the high rate of component performance improvements and thus system performance improvements and cost dramatically reduce the importance of preference overlap and heterogeneity.

Preference overlap and heterogeneity are partly analyzed using a modified form of Dosi's technological trajectory, which combines multiple trade-offs in a single figure. While an indifference curve represents the set of points for two or more variables to which individual or multiple users (e.g., a market segment) are indifferent, an innovation frontier[12] represents the trade-offs that multiple users (i.e., markets) and multiple designers make between different dimensions of performance and between performance and price. The direction of advance of these frontiers is a technological trajectory.

4

Computers

This chapter uses the history of computers to explore the time lag between identification of the concepts and architectures that form the basis of technological discontinuities and the commercialization of these discontinuities. It makes six related arguments about this time lag.

First, it shows how improvements in one type of component, vacuum tubes, made the first electronic computers (mainframes) possible in the 1950s. Second, it shows how following scientific and technological advances in transistors and integrated circuits (ICs) in the 1950s (more details are provided in Chapter 6) improvements in ICs drove improvements in computer performance and made a number of subsequent discontinuities possible. Third, while some of these discontinuities were initially adopted by new users, many of them would still have eventually emerged and diffused because they provided lower costs per processing speed (covered in Chapter 3) than did earlier computers and/or because they provided new benefits that most users desired. For example, minicomputers provided more user customization than mainframes, PCs and workstations provided faster response times than both minicomputers and mainframes, and portable computers such as laptops, PDAs (personal digital assistants), smart phones, and tablets provided more portability than PCs.

Fourth, computers had to exceed a minimum threshold of performance and/or fall below a maximum threshold of price before users would adopt them and, more specifically, their new discontinuities. This chapter focuses on minimum thresholds of performance for ICs that made these discontinuities possible. Fifth, the chapter shows how improvements in ICs changed the trade-offs that users and suppliers make between price and different dimensions of performance for computers. For example, improvements in processing speed that came from improvements in ICs caused the importance of processing speeds to eventually decline in comparison to response time.

Finally, the chapter shows how improvements in ICs had a greater impact on the diffusion of discontinuities than did heterogeneity or preference overlap. Although heterogeneity in user needs exists, as suggested by the initial adoption of low-performing discontinuities by new users, improvements in speed of many orders of magnitude have overwhelmed preference overlap/heterogeneity. This is partly argued using the concept of an innovation frontier, whose movements are driven by improvements in ICs and other components.

EARLY COMPUTERS AND COMPUTER SCIENCE[1]

The concept of a general-purpose computer was first characterized by Charles Babbage in the early to mid-19th century. Babbage wanted to develop a machine that could do mathematical calculations in order to eliminate the drudgery of manual calculations and reliance on error-ridden tables. With £1,500 in funding from the British government, he constructed a so-called difference engine consisting of wheels, gears, axles, springs, and pins that could compute 6-figure results. Unfortunately, its size (160 cubic feet), weight (15 tons), and complexity (25,000 parts and 400 square feet of drawings), along with a lack of improvement in these factors, severely limited the difference engine's potential.

Babbage then attempted another machine (called an analytical engine) based on the concept of the Jacquard loom invented by Joseph-Marie Jacquard in the early 1800s. This was a mechanical loom that used punch cards to control the sequence of weaving operations. The pattern of the loom's weave could be changed simply by preparing and inserting new cards, which can be thought of as a simple input-output device. Although Babbage was not able to develop such a "punch card" machine, Herman Hollerith did so in the late 19th century as high-tolerance mechanical parts became available. Punch-card systems saw extensive diffusion in the early 20th century partly because large firms were emerging that demanded methods for managing large amounts of information.

Simultaneous to these developments, others were characterizing concepts that later became important for computers. First, Babbage's colleague Ada Byron developed a process, a set of rules, and a sequence of operations that together would come to be called an algorithm. Second, George Boole developed the propositions of logic that would become the basis of so-called Boolean logic. This logic comprised a set of "AND," "OR," "IF," and other related statements that formed the basis for software code (and for electronic circuits).

Nevertheless, the ability to utilize these concepts did not appear until well into the 20th century, after the emergence of electricity, the vacuum tube, and the concept of stored program control. In the 1930s, Vannevar Bush constructed a mechanical device for solving second-order differential equations that was powered by electricity. At about the same time, Alan Turing and John von Neumann developed the concept of stored program control, where programs are stored in a computer and thus can be reused and modified. This concept was added to Babbage's computing machine as Turing, von Neumann, and others realized that the performance and costs of vacuum tubes had reached a point at which it was possible to construct an electronic version of it. In addition, during the late 1940s, many researchers worked on combining stored program control with algorithms and Boolean logic, and on characterizing possible computer architectures. While there have been many subsequent changes in the concepts associated with *electronic components* for computers (which are dealt with in Chapter 6), subsequent computers, including those defined as discontinuities, are all based on concepts and architectures that had been characterized by the late 1940s. By then engineers understood that they could design computers with different sizes of central processing unit (CPU), primary and secondary memory, word, and instruction sets, and that there would be various trade-offs among these design choices. For example, according to a computer scientist in the late 1980s,[2] "[computer designers had recognized that] much of computer architecture is unchanged since the late 1940s" and by the 1940s "architectural tricks could not lower the cost of a basic computer; low cost computing had to wait for low cost logic" (i.e., low-cost electronic components).

OVERVIEW OF DIFFERENT SYSTEMS[3]

Table 4.1 summarizes the five major technological discontinuities for electronic computers and classifies them as either radical or architectural innovations. Mainframe computers are considered a radical innovation because they involved a change from fixed to stored program control and this change also defined a new architecture. Minicomputers, PCs, and portable computers are considered architectural innovations because they are scaled-down versions of a mainframe. Workstations, too, are considered an architectural innovation because they use a smaller instruction set than do PCs, in what is known as reduced instruction set computing (RISC), along with faster microprocessors.

TABLE 4.1

Classification of selected discontinuities in computers

Discontinuity	Type	Date of introduction	Identification of concept or architecture (i.e., science)
Mainframe	Radical	Late 1940s	General computer: early 19th century Stored program control: 1936
Minicomputer	Architectural	1965	Late 1940s
PC	Architectural	1975	Late 1940s
Workstation	Architectural	1980s	Late 1940s
Handheld	Architectural	1995	Late 1940s

SOURCE: Moore (2004) and author's analysis.

Table 4.1 also compares the years of the emergence of the discontinuities with the years when the concepts and architectures for them were first identified and characterized. The basic concept was proposed by Charles Babbage in the early 1800s. It was partially implemented in the form of special-purpose mechanical computers that used punch cards in the late 1800s; this implementation can be considered the first "invention" of a computer, albeit only a partial one since it was not the general-purpose computer envisioned by Babbage. The mainframe computer is usually considered the first-general purpose computer. Subsequent computers, including those defined as architectural innovations, used architectures that had been characterized by the late 1940s. This means that the concepts and architectures that form the basis for all of these computers were known by that time.

Mainframe Computers[4]

Improvements in vacuum tubes enabled the creation of the first digital computers at British and U.S. universities in the mid-1940s and their commercialization in the early 1950s. These improvements had been driven by the market for radios since the early 1920s. By the 1930s they had begun to renew interest in Charles Babbage's computing machine and thus led to the development of stored program control and an architecture that supported its addition to Babbage's machine. The first customers for these electronic computers were well-established users of punch cards.

We can analyze the timing of mainframe computers in terms of minimum thresholds of performance and how improvements in vacuum tubes changed the design trade-offs for mainframes. In terms of a minimum threshold, the reliability of vacuum tubes finally reached a level at which a mainframe computer could operate for a "minimum number of minutes" before too many of the tubes "burned out."[5] In terms of trade-offs, as the cost of vacuum tubes dropped[6] and their reliability increased, the faster speeds of vacuum tube–based computers began to outweigh their disadvantages in higher costs and lower reliability when compared to the mechanical components in punch card systems.

Improvements in a broader set of components—in particular, magnetic components—also impacted the emergence of mainframe computers; thus this discontinuity involved larger changes in the linkages between components and depended more on a "novel combination of technologies" than did most of the other discontinuities discussed in this chapter. Improvements in magnetic components were driven by the market for magnetic tape players, which were first used in Germany in the 1930s and were applied to a wide variety of magnetic media, including "core," tape, drums, and disks; these were used for both primary and secondary memory. Different forms of magnetic media continued to be used in secondary memory long after transistors and ICs replaced them in primary memory.

Even with these improvements, however, not all potential users could afford mainframe computers even in the 1960s and 1970s. The high cost meant that only large organizations could utilize them to the extent necessary to justify purchase, and they used them in batch mode to achieve the necessary high utilization. Large organizations primarily leased their mainframes from suppliers such as IBM, and they depended on these suppliers for maintenance and access to their large libraries of software that came with a leasing agreement. On the other hand, small organizations accessed so-called remote computer services, such as processing and time sharing, through telephone lines, modems, and dumb terminals. Processing services included payroll, accounting, credit card, banking, financial, and health care, and were forerunners of the Internet services that later emerged. In the 1970s, however, historical accounts suggest that many firms believed that the diffusion of computer access would come about primarily through the diffusion of dumb terminals and remote services. Implicit in this belief was that remote access to large-scale computers through dumb terminals would continue to provide the most economical means of access.[7]

Minicomputers[8]

Improvements in ICs, which were primarily driven by military applications, led to the first minicomputer, Digital Equipment Corporation's (DEC's) PDP-8, in the mid-1960s. This was a scaled-down version of a mainframe in that it used a smaller CPU, a smaller primary memory, and a shorter word length and instruction set. The smaller CPU and primary memory, and the fact these circuits were made from standard as opposed to custom components, made the minicomputer much cheaper to produce than the smallest IBM System/360 on both an absolute and an instruction-per-second basis. For example, a user could rent the PDP-8 for about $525 a month, or 6 percent of the cost of IBM's smallest System/360, the Model 30.[9]

One reason that DEC and, later, other minicomputer manufacturers were able to sell a computer with such a small CPU was that they had found a new set of users and applications (see Table 4.2). Unlike accounting departments in large firms or government organizations, engineering departments wanted to use computers for various types of process control, engineering calculations, and other engineering applications. Focusing just on process control as an example, IBM did not offer such software as part of its leasing arrangements, and integrating a computer with factory processes required customization of its input-output devices. Although engineering departments were technically capable of doing this customization, they needed control over a computer, including access to accurate documentation. Unlike IBM, minicomputer manufacturers sold rather than leased their computers and provided users with the necessary documentation.

Although some might argue that this pattern conforms to Christensen's theory of disruptive innovation, the theory does not help us understand why minicomputers emerged and diffused when they did. A need for process control, engineering calculations, and other engineering applications had existed for many years and still exists in many large factories, so changes in demand are not the answer. A better explanation is that effective process control required computers whose performance exceeded a minimum threshold and whose price fell within the budgets (i.e., below a maximum threshold) of engineering departments. Since electronic components such as transistors and ICs had the largest impact on the performance and price of computers, we can also say that computerized process control required the performance of ICs to exceed a minimum threshold and their price to fall below a maximum threshold. Furthermore, since Moore's Law includes considerations of both price and performance, we can define the maximum threshold of performance in terms of the IC performance that existed when minicomputers were introduced in 1965 (see Figure 4.1).

Another way to look at the emergence of the minicomputer as a discontinuity is to analyze how improvements in ICs changed the trade-offs that computer users and suppliers were making between price and performance and between various performance dimensions. As prices fell and performance rose for ICs in the 1960s, the possibility of using standard components in smaller

TABLE 4.2

Changes in users, applications, sales channels, and methods of value capture

Discontinuity	Users	Applications	Sales channels	Method of value capture
Mainframe	Initially punch card users; later other users	Initially custom applications for military and other government agencies; later general business (accounting, inventory, logistics, payroll) and industry-specific software	Sales force	Leasing of computers and software
Minicomputer	Initially scientific and engineering companies; later other users	Initially engineering applications developed by users; later software for computer-aided design (CAD), manufacturing resource planning (MRP), and word processing	Initially mail order; later sales force	Sale of computers and software, together and separately; extensive documentation
PC	Individuals, universities, small businesses; later larger companies	Initially users (hackers) wrote software; later games and education followed by spreadsheet and word-processing software	Initially mail order; later retail, telephone, and Internet	Sale of computers and software, together and separately
Workstations	Design engineers	CAD	Initially sales force; later others	
Handheld	Different for personal digital assistants (PDAs) and "smart" phones	Different for personal digital assistants (PDAs) and "smart" phones	Retail and Internet	Different for PDAs and "smart" phones

SOURCE: Flamm (1988), Rifkin and Harrar (1988), Langlois (1992), Steffens (1994), Pugh and Aspray (1996), Ceruzzi (1998), and Campbell-Kelly (2003).

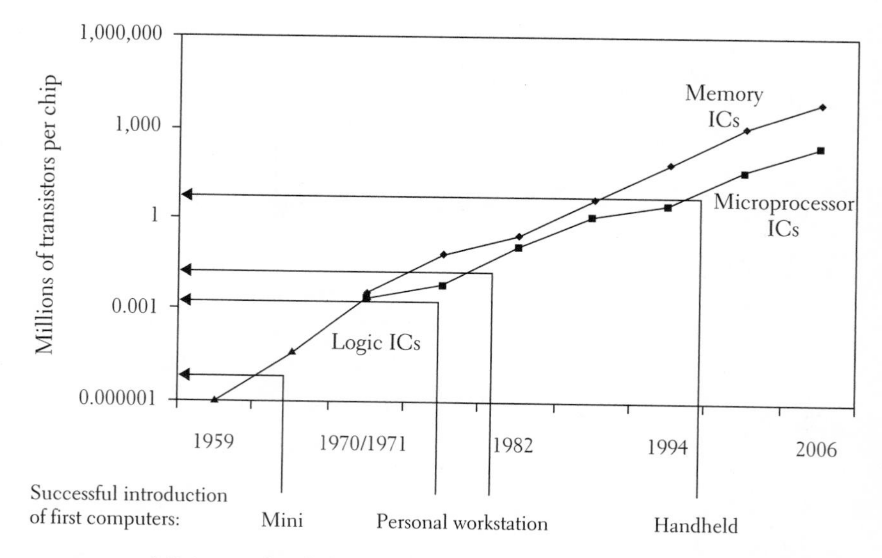

FIGURE 4.1 Minimum thresholds of performance in computer components
SOURCE: Moore (2004) and author's analysis.

and less expensive computers increased, as did the benefit-to-cost ratio of economical computerized process control, engineering calculations, and other engineering applications.[10]

Personal Computers[11]

By the mid-1970s improvements in ICs (and in magnetic recording density and cathode ray tubes) brought about the so-called microcomputer or PC, which was even smaller and cheaper (on an absolute and instruction-per-second basis), and less powerful, than the minicomputer. Because of reductions in feature size and thus increases in the number of transistors that could be placed on a chip, firms such as Intel (in 1971) were able to place a CPU on a single chip, called a microprocessor, which was initially used in calculators and aviation and scientific instruments. Improvements in microprocessors led to the first PC in 1975. PCs had slower processing speeds and shorter word lengths and instruction sets than did minicomputers, but faster response times and lower prices. Similarly, improvements in magnetic recording density made it possible for hard disk manufacturers to introduce smaller disk drives (5.25-inch and later 3.5-inch) that better matched the needs of PCs.

As with minicomputers, one reason that firms were able to sell a computer with such a small CPU, primary memory, and input-output device was that they found a new set of users and applications. Although many of these users already had remote access to mainframe or minicomputers, some wanted to write their own software and PCs were easy to obtain, customize, and modify. Others, such as recreational users, small-business owners, and university professors were interested in games, educational programs, word processing, and spreadsheets, for which fast response times were more important than the faster processing speeds of mainframes. While mainframes and minicomputers operated in batch mode, PCs provided faster feedback from each user keystroke and thus had faster response times for simple operations.

As with the minicomputer, some might argue that this pattern conforms to Christensen's theory of disruptive technology; however, this theory does not explain why PCs emerged and diffused when they did. A demand for game, education, word processing, and spreadsheet applications had existed for years, so it was not changes in demand that brought about PCs. A better explanation is that these applications required a computer with a minimum threshold of performance and a price that fell within a user's budget (i.e., below a maximum threshold). Since microprocessors and other ICs had the largest impact on the performance and price of computers generally, we can also say that using computers for these applications required microprocessors and other ICs to exceed a minimum threshold of performance. We can also say that by the mid-1970s the performance of microprocessors had reached the minimum threshold (see Figure 4.1) necessary to produce a computer that satisfied the minimum and maximum levels of performance and price for the PC's initial adopters.[12]

Another way to look at the emergence of the PC as a discontinuity and its gradual displacement of mainframes and minicomputers is to analyze how improvements in ICs changed the trade-offs that users and designers were making between price and performance and between various dimensions of performance. Because of their high cost, in the 1950s and 1960s, computers were used in batch mode in order to maximize utilization and thus minimize per-instruction cost. Better ICs reduced the costs and thus the importance of high utilization; this increased the potential importance of other dimensions of performance such as response time, which was also falling because of IC improvements. In hindsight, it is obvious that the importance of response time would increase as the cost of computing and thus the importance of high utilization fell. Individual control of a single computer, as with PCs, enabled faster feedback than with a mainframe or minicomputer as long as the processing requirements were not great, which was the case with the initial software applications mentioned earlier.

The data in Figure 4.2 supports the interpretation that the importance of processing speed fell as improvements in ICs occurred, thus driving changes in the trade-offs that users and suppliers of computers were making. It shows the innovation frontiers for computers from 1954 to 2002, plotting price versus processing speed for five different time periods. Each frontier represents the trade-offs between processing speed and price for a single time period; the movement of these frontiers over time represents the changing trade-offs that users and suppliers were making between them. We can say that improvements in electronic components pushed the frontiers toward the upper right portion of the figure—five orders of magnitude for speed and four orders for price. We can also say that these improvements caused the price of extra processing speed to decline and thus the relationship between price and processing speed to become flatter over time. This suggests that over time users placed less emphasis on processing speed than on price and other dimensions of performance (e.g., response time).

Furthermore, the interpretation in Figure 4.2 suggests that improvements in ICs (similar arguments can be made with magnetic recording density) had a much larger impact on the change from mainframes and minicomputers to PCs than did so-called preference overlap. While one theory is that the rate at which a discontinuity displaces an existing system depends on the degree of

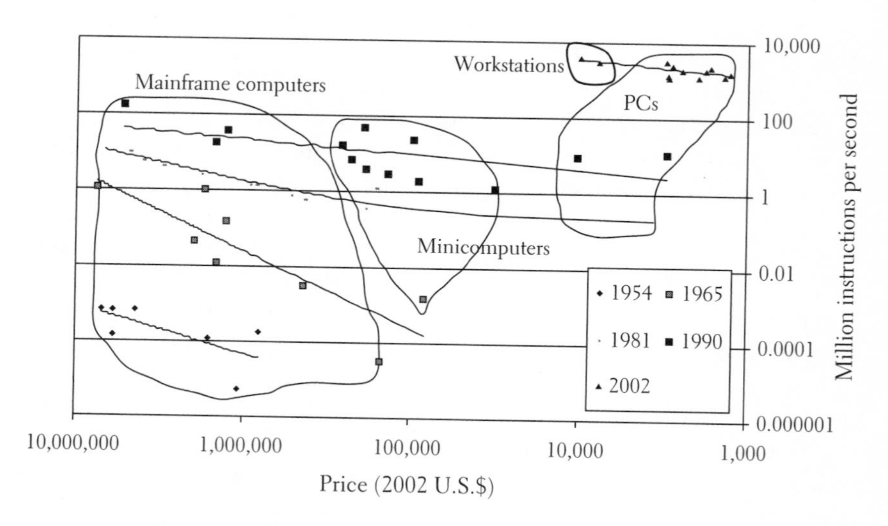

FIGURE 4.2 Innovation frontiers for computers

"preference overlap" between users,[13] the falling price of computer processing speed dramatically reduced this dependence. For example, although there were differences of one to two orders of magnitude in processing speed in 1965 among different computers, the five orders of magnitude in improvements between 1954 and 2002, which came from improvements in electronic components, made preference overlap insignificant.

Workstations[14]

Workstations are essentially high-end PCs that provide a higher level of performance. Like PCs, their introduction depended on improvements in microprocessors and other ICs in order for them to provide a minimum threshold of performance in various forms of engineering analysis and design. Although the minimum thresholds of performance were higher for workstations than for PCs, one way designers achieved them was by reducing the number of instructions in what is called reduced instruction set computing (RISC). The smaller instruction set enabled the computer's design to be optimized for a smaller number of instructions and, in combination with more expensive microprocessors, enabled workstations to provide better performance than PCs.

Another common aspect of workstations and PCs, which was not covered in the previous section, is the increasing use of "parallel processing." The falling costs of microprocessors led to an increase in a computer's microprocessors or so-called cores in a single microprocessor, which allowed the computer to perform more instructions in parallel. Although some historians consider this form of architecture to be significantly different from the "von Neumann architecture" described earlier, it is usually not considered a discontinuity since it was incrementally introduced in the form of gradually increasing processors, or cores in a processor, over many years. More important, the concept of parallel processing, first conceived in the late 1940s, was partially implemented in the 1950s and 1960s and was fully implemented only as the performance of microprocessors improved. Thus, parallel processing is completely consistent with this chapter's argument that inadequate components are the main reason for the time lag between the characterization of concepts or architectures and their commercialization.

Portable and Handheld Computers[15]

Improvements in microprocessors, ICs, and, to a lesser extent, hard disk recording density, batteries, and liquid crystal displays (LCDs) led to, and continue to lead to, new portable, including handheld, computers that are based on concepts and architectures that have been known since the late 1940s. The first technological discontinuity associated with portable computers was initially a failure. Although Osborne Computers introduced the first portable (better known as a "luggable") in 1981, the ICs available at that time did not meet the minimum thresholds of performance for processing speed and power consumption needed for compatibility with a desktop PC. Furthermore, unlike minicomputers and PCs, Osborne's portable was not able to find a new set of users who were not concerned with backward compatibility. The market for portable computers did not begin to grow until after improvements in a microprocessor's processing speed and power consumption, hard disk recording capacity, and LCD size and weight had reached minimum thresholds of performance such that a portable computer could satisfy users' minimum and maximum thresholds of performance and price.[16] As an aside, portable computers, like mainframes, depended more on a novel combination of components than did other discontinuities.

Similar improvements in components led to the emergence of handheld computers such as PDAs in the mid-1990s. Although the first PDAs were unsuccessful because of the high power consumption and low processing speeds of existing microprocessors,[17] a combination of improvements in microprocessors and reductions in functionality finally enabled a PDA to meet thresholds of performance and price for some set of users in 1995. Palm reduced the number of functions to a calendar, address book, and memo pad, which lowered the minimum thresholds of performance for the microprocessor, memory, and other components, and thus their costs.

However, further improvements in microprocessors, magnetic recording density, and both wireless and wireline systems continued to make new discontinuities possible. For example, they enabled Research in Motion (RIM) to include a phone receiver and a mail client in a PDA, resulting in the once ubiquitous BlackBerry. Similar improvements enabled the introduction of MP3 players and mobile phones that can play music, video, and other applications. Improvements in touch screen displays have also made new forms of computers possible, such as Apple's iPad.

THE FUTURE

Looking to the future, we see that continued improvements in ICs and telecommunication systems may end Microsoft's and Intel's ("Wintel's") hold on the PC industry or they may lead to completely new forms of computer networks. Improvements in ICs have eliminated the compatibility problems between Wintel and Apple computers, and improvements in hard disks and telecommunication systems have made it easier for users to store information on and access software over the Internet and thus bypass Microsoft Office. These improvements in "components" are changing the trade-offs that users make with respect to Wintel versus non-Wintel computers, desktop versus laptop, laptop versus tablet, packaged versus software as a service (SaaS), and on-site versus utility computing (see Chapter 9). In this way, they are driving the emergence of discontinuities.

Looking even further to the future, we see that improvements in ICs and in telecommunication systems may lead to new types of computers. Better ICs have made it possible to store product-related information on so-called RFID (radio frequency identification) tags. They may enable RFID tags and "smart dust" (a system of smart wireless sensors) to become decentralized networks of computers connected by small antennas and short-range wireless technologies.

CONCLUSIONS

This chapter used the history of computers to explore the time lag between the characterization of concepts or architectures and their commercialization. No matter how one defines the timing of these characterizations, there was a very long time lag before the commercialization of general-purpose computers. If Charles Babbage's work can be defined as a sufficient characterization of the concept of a computer, it took more than 100 years for components such as vacuum tubes and transistors to emerge and make his machine economically viable. Even if we define this characterization in terms of a proof of concept, we can say that since the first special-purpose computer (i.e., the punch card system) was invented in the late 1800s, the time lag between special-purpose computers and general-purpose computers was about 60 years. If we focus on when the architectures for electronic computers were understood and define this as 1950, the time lags were about 15 years for minicomputer (1965), 25 years for PCs (1975), and 45 years for handhelds (1995), where each of these discontinuities can be defined as a new industry.

As noted previously, one theory is that the primary reason for this time lag is a need for complementary technologies and science that sometimes depend on novel combinations of components or technologies. This chapter showed that in the computer industry, the time lag was primarily due to the lack of electronic components. Charles Babbage's computer had to wait more than 100 years for components such as vacuum tubes and transistors to make it technically and economically viable. Following a second round of scientific advances in the 1950s for transistors and ICs, continued improvements in the number of transistors on a chip enabled the subsequent commercialization and diffusion of technological discontinuities such as the minicomputer, PC, workstation, and portable (including hand-held) computers, which are all based on concepts and architectures that were known by the late 1940s.

However, this does not mean that research on computer design or novel combinations of components has not had an impact on competition between firms. Many observers have described how firms such as IBM, DEC, Microsoft, Intel, Dell, and others conducted research and made good and bad design decisions, including those that involved novel combinations of components and finding new customers, and how these decisions affected the performance and price of their products and thus competition in the industry. Furthermore, the success of specific products defined as technological discontinuities also depended on design decisions that might be seen as novel combinations of components. The point here is that these design decisions had a much smaller impact on a computer's *long-term* improvements in processing speed, response time, and portability, and thus on the timing of the discontinuities or their diffusion, than did improvements in electronic components. Even if these design decisions doubled, tripled, or even quadrupled performance or the performance-to-price ratio of computers, these improvements were very small compared to the improvements of five orders of magnitude in speed and four orders of magnitude in price (see Figure 4.2) that came from ten orders of magnitude in improvements in ICs (see Figure 4.1).

One way to analyze the timing of system-based discontinuities is to look at the trade-offs that both users and suppliers make between price and performance and between different dimensions of system performance, and to analyze the impact of improvements in components on these trade-offs. Over the history of computers, users and designers have made trade-offs between price, processing speed, ease of customization, response time, portability, and other dimensions of performance. Initially, because price and processing speed were the most important variables, computers operated in batch mode and only large organizations could afford them. However, as IC improvements drove gains in both processing speed and price, users and designers began to place more emphasis on factors such as ease of customization in minicomputers, response times in PCs, and low weight in portables. While Christensen characterizes these changes in emphasis in terms of "technology overshoot," this book uses the term *trade-offs* because it can be applied to both low-end and high-end innovations and because both users and designers make these trade-offs. For example, improvements in ICs changed the trade-offs that users were making between owning a computer and accessing one remotely through dumb terminals and, in doing so, drove the diffusion of PCs where defining the PC as a high-end or a low-end innovation when compared to dumb terminals is highly problematic.

A second way to analyze the timing of system-based discontinuities is by considering the minimum required threshold of component performance before the system-based discontinuity becomes technically and economically feasible. A minimum level of performance was needed in vacuum tubes before mainframe computers became feasible and in ICs before minicomputers, PCs, and several portables did so. Since Moore's Law includes not just performance but also price in its characterization of ICs (see Chapter 6 for more details), this chapter used data on Moore's Law to define minimum thresholds of IC performance for the first minicomputers, PCs, and portables.

Furthermore, the demand for a new discontinuity such as PCs had a relatively small impact on IC performance; thus the theory about demand driving reductions in cost is not very useful for understanding the emergence or diffusion of discontinuities in computers. Instead, improvements in ICs, which were driven by existing computers and other electronic systems, gradually improved the "potential" cost and performance of PCs, and growing demand for PCs only affected the cost of ICs as they were modified for PCs, as well as mechanical assemblies such as keyboards. For example, increasing demand for PCs encouraged IC manufacturers such as Intel to customize their microprocessors for them. However, after these components were customized, reductions in feature size that occurred as part of Moore's Law probably had a larger impact on the rate of improvement in these microprocessors and other ICs than did PC demand.

In summary, this chapter helps managers and students analyze the timing and the diffusion of potential technological discontinuities. For example, the fact that ICs had a strong impact on computer performance suggests that firms should have been focused on the rate of IC improvement when they considered potential discontinuities. The fact that computers and ICs also had high development costs suggests that firms should have looked for discontinuities that would have lower development costs. The fact that smaller computers such as PCs had lower costs per processing speed and that everyone wanted fast response times suggests that the chances for their diffusion were high.[18] Although this might be called 20-20 hindsight, these are some of the questions that managers and students should address when they analyze potential discontinuities.

5

Audio and Video Recording and Playback Equipment[1]

This chapter uses the history of audio and video recording and playback equipment, with a focus on magnetic tape, to further explore the time lag between the identification of the concepts and architectures that form the basis of technological discontinuities and the commercialization of these discontinuities. It makes six related arguments about this time lag.

First, as with computers, improvements in components drove the emergence of discontinuities. For magnetic tape systems, it was improvements in the recording density of tape, which benefited from reductions in the scale of magnetic storage regions (i.e., geometrical scaling). The resulting improvements drove the emergence and performance of all tape-related discontinuities, partly because they enabled the replacement of complex designs with relatively simple ones. The simplest designs had a single stationary recording head while the more complex had nonstationary, and sometimes multiple, recording heads. Fewer, stationary, recording heads are less expensive but store less data.

Second, this chapter defines the minimum thresholds of performance for magnetic recording density that made these discontinuities possible. A minimum threshold of performance was needed before new magnetic tape systems could exceed it and/or fall below a maximum threshold of price for new users and thus cause them to adopt the simpler and cheaper designs.

Third, the chapter shows how improvements in recording density changed the trade-offs that users and suppliers made between price and different dimensions of performance; these changes facilitated the emergence and diffusion of the discontinuities. In particular, improvements in recording density lessened the importance of audio and video quality and increased the importance of price and small size.

Fourth, improvements in magnetic recording density had a larger impact on the diffusion of the discontinuities than did the degree of preference overlap/heterogeneity. This is partly argued using the concept of an innovation frontier, movements in which were driven by improvements in recording density and in other components.

EARLY FORMS OF MAGNETIC SYSTEMS[2]

The first idea for magnetic recording built on Faraday's Law and its application to telephones and mechanical recording devices. Sound vibrations were recorded on a magnetic wire with a magnetic coil, and the wire was moved between two spools. Although this device was first used for dictation by Vlademar Poulsen and others in the early 20th century, the lower cost of tin foil, wax, and later phonograph recorders, which were only marginally successful for dictation,[3] prevented much growth in the market for wire recorders. It was not until the 1930s, in Germany, that improvements made magnetic recording and playback possible. Following a cigarette manufacturer's successful efforts to coat paper with magnetic material and build a prototype in 1931, BASF began research on a variety of magnetic and related materials and AEG developed the "magnetic ring head" that became the basis for subsequent recording heads. The initial magnetic material, carbonyl iron, was replaced by magnetite and gamma ferric oxide tape in the 1930s and by chromium dioxide in the 1970s. The magnetic material was initially deposited on paper and later on cellulose acetate (by the 1930s) and plastic (by the 1940s). The development of these materials and processes for attaching them to base materials also involved advances in science that formed the initial base of knowledge for magnetic recording and playback equipment.

System-related advances also occurred in the 1940s and 1950s. First, the ability to reduce noise and distortion through AC bias was developed independently in Japan in 1938 and in Germany in 1940. Because tape has a nonlinear response at low signals, adding an inaudible frequency signal to the audio signal corrected AC bias and thus reduced noise and distortion. Second, as mentioned previously (and discussed in more detail in later sections), engineers in the United States found that a single recording head could not provide sufficient recording capacity and so in the 1950s they developed more complex systems that had multiple recording heads, some of which were not stationary. As an aside, the first hard disks were also developed in the 1950s.

By the end of the 1950s, then, most of the concepts and architectures that formed the basis of magnetic recording and playback systems had been characterized, as had the concepts that formed the basis of digital recording such as pulse code modulation (PCM) and differential PCM, which had been developed much earlier. The exceptions are Dolby Sound and the concepts that formed the basis of optical discs and digital compression. Dolby Sound was developed in the 1960s, as were the first semiconductor lasers and photodiodes, which made optical discs possible. Compression techniques are still being developed in response to improvements in ICs, which are also a key component in magnetic recording and playback systems.

OVERVIEW OF DIFFERENT SYSTEMS[4]

Most audio and video recording and playback equipment requires components such as a recording medium, a read-write mechanism, input-output devices, and, typically, some form of amplifier. Magnetic tape or hard disks are the media, and a magnetic head performs the read-write operations. For magnetic tape, single or multiple heads can be used where each can be stationary or rotary (i.e., it moves across the tape). Single stationary heads are the simplest and thus the smallest and cheapest. Optical recording equipment uses metallic disks as the media and a laser and photodiode to perform the read-write operations. For input-output components, audio equipment uses microphones and speakers while video equipment uses cameras and displays. The devices may be included in the equipment or attached to it such as in the case of televisions and video players. Vacuum tubes, transistors, and ICs have been used for amplification, and now flash memory, a form of IC, is used in place of magnetic disks and optical discs.

Table 5.1 summarizes the technological discontinuities for both audio and video equipment and the changes in design, initial customers, and applications for them. Changes in design include conversion from analog to digital and from magnetic tapes to disks; several smaller yet critical changes within magnetic tape; changes between single stationary, single rotary, and multiple rotary magnetic heads; and recent conversions to optical and flash memory. Stationary heads do not move while the tape passes by them, while rotary heads move as the tape moves in order to access and store data on a greater portion of the tape's width. Major users changed from broadcasting and music firms to final consumers. The initial applications were different for many discontinuities and included editing by broadcasting and music firms, and playback and recording for consumers.

TABLE 5.1

Technological discontinuities in recording and playback equipment

Discontinuity	First decade of use	Key aspects of product design and the changes in them	Initial customers and applications
Analog audio		*Magnetic wire/tape passes a magnetic head or coil*	
Wire	1880s	Wire on a spool	Dictation
Reel-to-reel tape	1940s	New material (tape); single stationary head in open reel	Prerecorded radio broadcasts
8-Track	1960s	Smaller tape with enclosed	Prerecorded music for
Cassette tape	1960s	reels and fixed threading	car/portable users
Analog video		*Magnetic tape and head*	
Quadruplex	1950s	Four rotary heads	Prerecorded television broadcasts
Helical scan	1960s	Single rotary head	Education and training
Camcorder	1970s	Added camera and display	News gathering
Digital audio		*Various methods*	
Hard disk	1980s	Magnetic head and disk	Editing by music firms,
DASH	1980s	Replaced disk with tape	broadcasters
Optical discs	1980s	New disk (metal), read/write method (lasers/photodiodes)	Prerecorded music
DAT	1980s	Returned to magnetic head and tape but with rotary head	Editing by music firms
DCC	1990s	Replaced rotary with stationary head	
Flash memory	2000s	All done by integrated circuits	Portable devices
Digital video		*Various methods*	
Tape	1980s	Magnetic head and tape	Editing by television
Hard disks	1980s	Replaced tape with disk	broadcasters
Optical discs	1990s	New disk (metal), read/write method (lasers/photodiodes)	Prerecorded movies
Flash memory	2000s	All done by integrated circuits	Camcorders, phones

SOURCE: Inglis (1991), Millard (1995), Sadashige (1999), Sugaya (1999), Grindley (1995), and various web pages.

NOTE: DASH, digital audio stationary head; DAT, digital audio tape; DCC, digital compact cassette.

Key dimensions of performance for recording and playback equipment are recording time, sound and image quality, size, and editing capability. Table 5.2 summarizes the design decisions that impacted these dimensions and thus the trade-offs between design decisions. For example, although both wider and longer tapes and slower tape speeds increase recording time, the former also increase costs and the latter reduce sound or image quality. Also, while faster tape speeds, multiple recording heads, and rotary as opposed to stationary recording heads can process more data and thus produce higher sound and

TABLE 5.2

Impact of design-related decisions on key dimensions of performance

Dimension	Design decision
Recording time	Tape width, tape length, tape speed, quality
Sound or image quality	Amount of data processed and data digitalization (reduced error loss)
Editing capability	Data digitalization
Size	Tape width, length, speed

SOURCE: Inglis (1991), Millard (1995), Sadashige (1999), Sugaya (1999), Grindley (1995), and various web pages.

image quality, they also increase costs through their need for wider and longer tapes and more complex tape-handling systems. Digital systems, like video, also require more data processing, and so, although they have less data loss and better editing capability than do analog systems, their greater data requirements can increase costs through the need for wider and longer tapes, more complex tape-handling systems, and more expensive ICs. Data compression is one way to reduce the differences in data requirements between analog and digital, and among compression techniques there is a trade-off between reducing the amount of data and lowering the quality of the sound or image.

Fixing the recording time, which is ordinarily done with packaged music and video, enables one to focus on the trade-offs between quality and size and between quality and price; these trade-offs can also be used to define innovation frontiers for magnetic recording and playback equipment. Using data on actual systems (see the Appendix for more information), Figures 5.1 and 5.2 show the trade-offs between quality and size and between quality and price, respectively. They also show how improvements in magnetic recording density pushed the curves toward the upper right of the figure and thus changed the trade-offs between performance and price.[5]

The following sections address all of the discontinuities (not just tape-based) in more detail. Each discontinuity is described in terms of (1) changes in design and applications; (2) changes in performance and price; (3) improvements in components and the industries that drove the improvements; and (4) relationships among these changes and improvements.

Analog Audio

As discussed previously, the concepts that formed the basis of stationary-head magnetic tape systems had been characterized by 1940. Engineers basically understood the basic science of magnetic materials: how they attached to base materials such as cellulose acetate and later polyvinyl chloride (PVC) plastic;

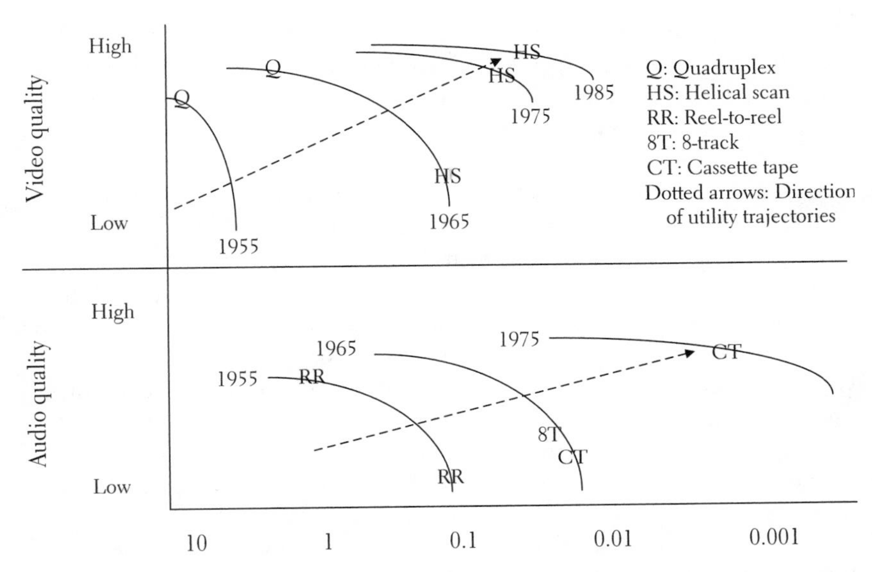

FIGURE 5.1 Innovation frontiers for audio and video equipment (quality versus size)

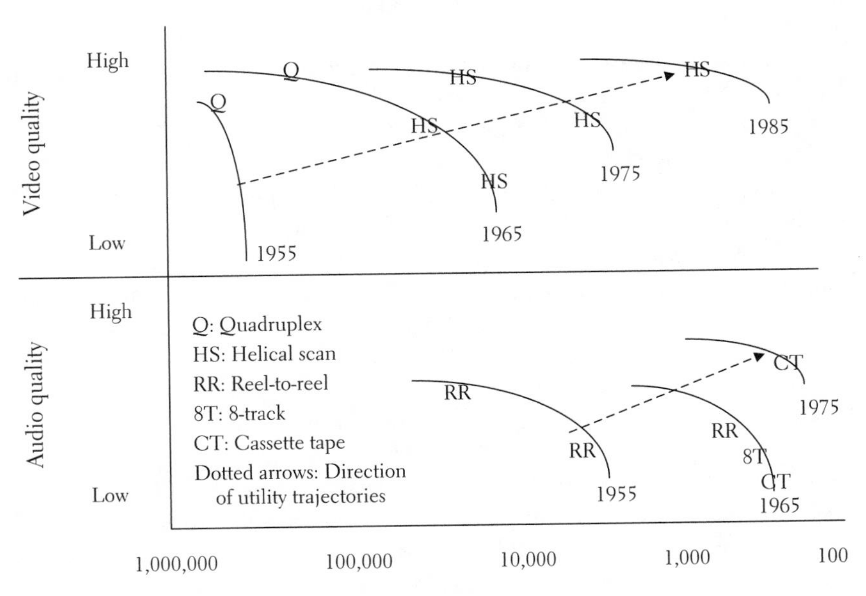

FIGURE 5.2 Innovation frontiers for audio and video equipment (quality versus price)

the concept of AC bias; and how tape could be stored, like film, on large reels. Instead, the commercialization and diffusion of the first analog systems depended on improvements in components such as magnetic coils, PVC plastics, and vacuum tubes, and on finding an appropriate market. These improvements were driven by the market for telecommunication equipment (Engel 1999), many consumer products, and radios and music players, respectively.[6]

Reel-to-reel magnetic tape was sold to a new set of users. Although the first magnetic tape players were used by the German government in the 1930s to support mass rallies and other aspects of Germany's propaganda machine, it was not until after World War II that improvements enabled tape players to be used in the first successful commercial application, prerecorded radio broadcasts in the United States. The ability to prerecord radio programs reduced programming costs and eased the work schedule of popular entertainers such as Bing Crosby. By the early 1950s, magnetic tape was used by most radio broadcasters for time-delayed broadcasts and by music companies for editing music before they created a "record master."[7]

In the 1960s, the smaller size of tape and its lower sensitivity to vibrations than that of phonograph records opened up a market for portable tape players, first for automobiles and later for handheld players. One reason this market emerged was that improvements in the magnetic recording density of tape and, to a lesser extent, in the strength and cost of plastics enabled suppliers to use a slower tape speed, a shorter reel of tape, and, in the case of cassettes, a thinner tape than was used in reel-to-reel players. These changes reduced the size of the tape player by more than 50 percent primarily by simplifying the tape-handling system. Improvements in the cost and strength of plastics were driven by a number of consumer products and contributed to reductions in tape player size.[8] Another reason that firms were able to reduce size was that they partly sacrificed sound quality, particularly in cassette recorders; for 8-track systems they eliminated the recording function. These changes came about because manufacturers were targeting new market segments such as portable players in automobiles with 8-track players and the low-end recording market with cassette recorders.[9]

Continued improvement in magnetic recording density gradually eliminated the sound quality disadvantages of these portable players as compared to reel-to-reel machines. By the mid-1970s cassette tapes had largely replaced both reel-to-reel tape and 8-track cartridges, and, along with improvements in other components such as ICs, improvements in recording density led to Sony's release of the Walkman in the late 1970s.[10] Although some might argue that this pattern conforms to Christensen's theory of disruptive technologies, this theory does not address how the performance of the first 8-track and

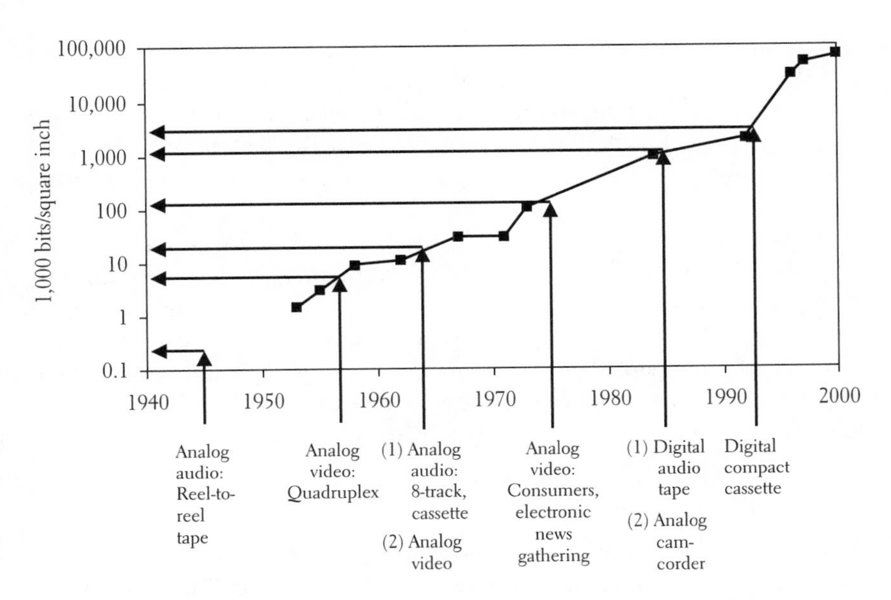

FIGURE 5.3 Minimum thresholds of performance for magnetic tape in systems

cassette players had reached a "minimum threshold" or how their prices had fallen below a "maximum threshold" that would enable their diffusion. A better argument is that by the early 1960s recording density had reached the minimum threshold of performance (see Figure 5.3) necessary for a shorter and thinner tape (in the case of cassettes) and a slower tape speed; these changes enabled the use of a simpler and thus cheaper tape-handling system.

Analog Video[11]

Incremental improvements in the magnetic recording density of tape and, to a lesser extent, in transistors and ICs made the first analog video recorders possible and led to several technological discontinuities within them (see Table 5.1). During the 1960s and 1970s, improvements in recording density were driven by the market for audio tape recorders and computers while improvements in transistors and ICs were driven by demand initially for military applications and later for computer applications. Improved magnetic recording density, transistors, and ICs were particularly important because video requires about 100 times the amount of data that audio requires. Although eventually recording density improvements made it possible to handle this large amount of data with systems that were similar to audio systems, initially such "stationary-head reel-to-reel" machines required very fast tape speeds and thus excessively complex tape-handling capabilities. For example, even after several years of trying to apply a stationary head system to video recording (often called the longitudinal approach), in 1953 RCA's system still required tape speeds of 360 inches per second and a 17-inch reel to record four minutes of playing time.[12]

Ampex was the first manufacturer to create a design that could effectively handle the higher data requirements of video recording. Named for its four rotating heads that simultaneously carried out read-write operations on different parts of the tape, the Quadruplex required tape speeds of only 15 inches per second and thus enabled a simpler tape-handling system whose benefits outweighed the extra cost of multiple heads. Ampex demonstrated the system to television broadcasting executives in early 1956 and began deliveries in 1957.

Videocassette Recorders

Incremental improvements in magnetic recording density continued during the 1960s and 1970s and, albeit with some sacrifices in quality, led to the introduction and diffusion of simpler systems than the Quadruplex. Sacrifices in quality were made as new applications such as training and education and new customers such as sports teams, airlines, and universities were targeted. The most successful of the new single-rotating-head designs was known as the helical-scan recorder because of the way the tape wrapped around a rotating head in a helix.

As documented by many others,[13] helical-scan systems were introduced in the early 1960s. By 1971 several Japanese firms, including Sony, Panasonic, and JVC, were selling compatible products based on the U-Format standard. However, Sony was unable to convince Panasonic and JVC to introduce products that were compatible with its updated version of the U-Format, which it called Betamax. Instead, Panasonic and JVC introduced products that were based on an alternative standard, VHS, which became the standard for video recording and playback machines or, more commonly now, videocassette recorders (VCRs). Furthermore, recording density improvements gradually eliminated the quality disadvantages of helical-scan systems, which caused television broadcasters to begin using them in place of the Quadruplex in the early 1980s.

Again, the story of videocassette recorders may sound like Christensen's theory of disruptive innovation, but this theory does not address how the first helical-scan system's performance reached a "minimum threshold" and its price fell below a "maximum threshold." A better explanation is that by the mid-1960s recording density had reached the minimum threshold of performance for a "component" (see Figure 5.3) that was necessary to achieve minimum and maximum thresholds for the tape-based "system." The higher density enabled firms to use fewer recording heads, slower tape speeds, and thinner tape; these changes also enabled the use of a simpler, smaller, and cheaper tape-handling system. A question is why helical-scan system replaced the dominant Quadruplex. Although one theory is that this was the result of low preference overlap, there does not appear to have been low preference overlap between the users of VCRs and the Quadruplex. As with computers, a better explanation is that dramatic improvements in recording density led to improvements in size, quality, and price, and thus dramatically reduced the importance of preference overlap between different users. For example, the size reductions of three orders of magnitude (see Figures 5.1 and 5.2) for both audio and video equipment over about 40 years were much greater than the differences of one to two orders of magnitude that existed among systems in a specific year.

Camcorders[14]

The combination of cameras and recorders (and thus the term *camcorder*) in a single product makes this discontinuity more complex and more dependent on a "novel combination of components" than previous discontinuities. The first camcorders were used by news organizations in the 1970s for so-called electronic news gathering because they enabled faster editing than with 35mm photographic film. Reductions in size and price drove their diffusion in electronic news gathering, and these improvements and the introduction of systems that sacrificed quality for cost enabled their gradual diffusion among consumers starting in the mid-1980s.

The key drivers of improvements in camcorders came from improvements in cameras, recording density of tape, and ICs. A picture of one system from 1976[15] suggests that the camera represented about one-half the product space and the VHS recorder, which resembled the home units of the time, occupied the other half. Significant reductions in camera size came from the development of charge-coupled devices (CCDs), which were first used in facsimiles; their improvements were driven by factors similar to those that drove improvements in ICs (see Chapter 6).

Improvements in the recorder portion of the camcorder were driven by the same set of issues covered in previous sections, along with shorter tape reels and the move from 0.5-inch to 0.32-inch (8-mm) tape, which was not compatible with VHS or Betamax. The use of shorter and smaller tape reduced the size of the cartridge and simplified tape handling; thus it enabled reductions in both size and cost. One reason that shorter and smaller tape could be used was that image quality (and recording time) was initially sacrificed to produce a smaller and cheaper camcorder. A second reason was that improvements in magnetic recording density eventually eliminated most of the quality and recording time disadvantages. A third reason was that improvements in ICs enabled firms to place a VHS player mechanism and circuitry in a camcorder for a little more money and a small increase in physical size, so that 8-mm tape could be played on a home VCR.

Digital Audio[16]

Incremental improvements in the magnetic recording density of tape, in ICs such as microprocessors, and in other semiconductors such as lasers and photodiodes have led to the emergence of several new technological discontinuities that can be classified as digital audio systems. When an analog signal is converted to a binary form, it is called a pulse-code-modulated signal. This technology was developed by telephone companies in the 1930s in an attempt to increase the capacity of their telephone lines. The digitalization of audio signals can reduce data loss, enable error correction, and facilitate editing. The disadvantage of digital audio is that even with the use of data compression, it requires about 30 times more data and thus longer tape and more transistors than analog audio systems require. Methods of data compression such as MPEG-3 may be the only recent scientific development that has impacted the use of digital audio. Instead, improvements in magnetic recording density, ICs, and other components have enabled tape-based digital audio systems to become technologically possible, albeit not commercially successful, particularly among consumers.

The largest applications for digital audio systems have been those for music companies and radio broadcasters. These applications use a variety of storage technologies within a so-called storage hierarchy, where access times are more important than cost at the top and costs are more important than access times at the bottom. By the late 1990s, improvements in recording density of hard disks and processing speed caused the PC to become the interface between recording engineers and digital synthesizers; the MIDI (musical instrument digital interface) sequencer to become the word processor for music, and various disks (e.g., 3.5-inch); and USB devices to become the media for transferring data between PCs, digital synthesizers, and other equipment.

At the bottom of a music company's or radio broadcaster's hierarchy of storage systems was digital tape, used as backup for music and radio programs. The first digital tape systems used stationary heads like those in cassette systems. The most successful of these was the digital audio stationary head (DASH).

However, the large amounts of data that had to be stored in a digital format required very fast tape speeds and thus, as in the case of analog video systems for television broadcasters, digital tape was not widely used until a more complex tape system, in this case a rotary head system called digital audio tape (DAT), was developed in the early 1980s. Further improvements in recording density finally enabled a return to stationary heads in the 1990s with a new format called digital compact cassette (DCC). As discussed in the next section, both DAT and DCC were introduced in consumer products, albeit unsuccessfully.

The first successful digital recording medium for consumers was the optical disc, which is better known as the compact disc (CD). CDs provided higher quality and smaller size than analog tape and for a reasonable price. As summarized in Chapter 2, laser beams record bits of information on a coated disk by thermally heating very small areas on it. Heat changes the reflectivity of these areas, which can be sensed by a combination of a laser and a photodiode. Improvements in optical discs (covered in Chapter 2), along with improvements in microprocessors and other ICs, gradually improved the price and performance of CDs and thus made them economically feasible.

As for tape-based digital systems, DAT and DCC have been much less successful than the CD in consumer applications. Their performance advantages over CDs are small and music companies did not strongly support them. Moreover, DAT was not backward compatible with cassette tapes (albeit DCC was), and the network effects associated with CDs may have been too large for DCC. Finally, firms have not found new consumer applications. Nevertheless, the only reason that manufacturers attempted to target consumers with various digital audio systems (and partly why music companies use DCC) was the incremental improvements in magnetic recording density, which made it possible to develop such systems for a price and level of performance that was *thought* to be potentially suitable for consumers.

Looking more broadly at the digitalization of music, improvements in microprocessors, magnetic disks, and flash memory have played a role in the diffusion of the Internet and are now impacting the downloading of music. The PC is becoming the main music player in the home, where, as in music companies, hard disks have become the main form of storage. In portable devices, Apple's iPod Nano and mobile phones use flash memory to store music. Continued improvements in the capacity and price of flash memory and other chips will probably cause mobile phones to become the dominant portable music player.

Digital Video[17]

There are a number of similarities between the concepts of digital audio and video, their advantages and disadvantages when compared to analog systems, the improved components that drove their emergence, and their initial applications. First, digital video is based on a form of pulse code modulation, differential pulse-code modulation, which only stores the digital data that is different from the previous frame. Second, like digital audio, digital video can reduce data loss, enable error correction, and facilitate editing. Once an image has been transformed into 1s and 0s, it is easy to alter its size, aspect ratio, brightness, color, and linearity. Third, digital video requires much more data than analog video. Fourth, improvements in the recording density of both magnetic and optical media and the use of data compression techniques, which were made possible by improvements in ICs, have enabled commercial systems to emerge and diffuse.

Fifth, like digital audio, digital video was first used in professional applications and those at the bottom of a "hierarchy" of storage systems where cost is more important than access time. In the 1990s television broadcasters used hard disks in applications that required short access times and tape or optical-based systems, where cost was more important than access time, for the so-called backup of data. In the late 1990s, television broadcasters began replacing analog-based tape systems with digital systems for data backup and for electronic news gathering. While initially raw news footage was collected with analog-based camcorders and transferred to a magnetic hard disk on a PC with an appropriate reader, improvements in recording density enabled digital-based tape systems to replace analog-based systems in these camcorders.[18]

Similar improvements in magnetic tape and ICs enabled the digitalization of camcorders for consumers. Although magnetic tape was used in some digital camcorders during the mid-1990s, incremental improvements in flash memory finally reached the point at which digital camcorders' advantages in size and access times outweighed their disadvantages in cost and storage capacity. A plot of storage capacity versus access time, or of storage capacity versus price, for flash memory would reveal the same type of curve shown in Figures 5.1 and 5.2. Not only would the innovation frontier be moving in the direction of increasing storage capacity and lower price; it would show that the price difference between flash memory and tape has declined substantially. We can think of the price that existed for flash memory–based systems when they began replacing tape-based systems as their maximum threshold of component price. Similar arguments can be made for camera phones and video phones.

The other major consumer product that uses digital video technology is of course the digital video disc (DVD). Improvements in lasers, rotation speeds, error correction codes, and servo systems have changed the design trade-offs for video recording systems, leading to two technological discontinuities, of which the most recent is Blu-ray. Shorter-wavelength lasers and a higher numerical aperture lens have reduced the size of the memory spot, and improvements in ICs have enabled more powerful error correction codes and faster servo systems. In the ten years that followed the introduction of the first CD, these improvements increased CD capacity by more than 10 times, reduced access times by two-thirds, and increased transfer rates a 1000 times. Of course, the life of DVDs may be just as short as the life of CDs appears to be. The ability to download music from the Internet and the greater portability of semiconductor memory is eliminating the need for CDs, and similar trends, noticed more than ten years ago,[19] may eliminate the need for DVDs.

THE FUTURE

The future for magnetic tape or optical disc storage does not look very bright. The problem is that the technology paradigm for magnetic tape does not allow reductions in size or increases in speed to the same extent achieved by the technology paradigm of ICs. As with optical and magnetic disks, the size of the read-write head prevents magnetic tape systems from ever being made as small as an IC; this problem also prevents access speeds from ever reaching those of ICs. Although the cost per memory bit is still much lower with tape than with ICs or even hard disks,[20] most users prefer the higher speeds of disk drives for desktop applications and the smaller IC sizes for portable applications. Magnetic tape has been replaced by flash memory in video cameras and portable music players and by optical discs in video playback machines. It is likely that optical discs will also eventually be replaced by ICs as IC improvements continue and as more video is downloaded from the Internet rather than sold in packaged DVDs.

Nevertheless, improvements in the magnetic recording density of tape continue to be made. For example, IBM announced in January 2010 that its scientists had achieved a recording density of 29.5 billion bits per square inch—about 39 times the recording density of existing commercial products at that time[21] and about 300 times the recording density achieved in 2002 (see Figure 5.3). Thus the limits to geometrical scaling discussed in Chapter 3 have not yet been reached, and these improvements continue to enable magnetic tape to be used for reliable backup storage.

CONCLUSIONS

This chapter used the history of audio and video recording and playback equipment, with a focus on magnetic tape, to explore the time lag between the characterization of concepts or architectures and their commercialization. Since all of the concepts and architectures for tape-based systems had been identified by the late 1950s, this time lag ranged from less than 10 years for 8-track and cassette equipment to about 40 years for digital audio players such as DCC. And, since digital video players based on magnetic tape never diffused among consumers, we can say that the time lag for them was even longer than it was for analog audio players. Moreover, since this time lag was largely a function of improvements in one component, the magnetic recording density of tape, it depended more on improvements in this component than on "novel combinations of components."

This does not mean that research on recording and playback equipment or novel combinations of components has not affected competition between firms. As in the computer industry, the management literature on recording and playback equipment emphasizes how incumbents such as Sony, Philips, and others conducted research and made good and bad design decisions, including those that involved novel combinations of components and finding new users, and how their decisions impacted the success of these incumbents in all of the discontinuities until the emergence of the MP3 player. Furthermore, the success of specific products defined as technological discontinuities depended on design decisions that might be defined as novel combinations of components. The point here is that these decisions had a much smaller impact on *long-term* improvements in the performance of magnetic storage systems, and thus the timing of discontinuities, than did improvements in magnetic recording density. Improvements of about six orders of magnitude in density enabled reductions of four orders of magnitude in price and size for tape-based systems.

One way to analyze the timing of system-based discontinuities is to look at the trade-offs that both users and suppliers make between price and performance and between different dimensions of performance for a system, and to determine the impact of component improvements on these trade-offs. Over the history of recording and playback equipment, users made trade-offs between price, quality, size, recording time, and other performance dimensions. For magnetic tape–based systems, improvements in recording density were used to increase some combination of these variables. Since recording time was fixed for packaged music and video, these improvements could be used to improve quality or size; the resulting improvements in audio and video quality eventually led to a lessening of their importance. On the other hand, for optical discs the improvements in lasers, photodiodes, and metal coating led to increases in quality that were so much greater than those for magnetic tape and phonograph records (and size) that they caused users to emphasize quality over price. Optical discs for both audio and video playback are a good illustration of how characterizing these changes in terms of trade-offs instead of in terms of Christensen's "technology overshoot" enables one to analyze both high- and low-end innovations. In both cases, improvements of many orders of magnitude in magnetic and optical recording density probably had a larger impact on diffusion of discontinuities than did the degree of preference overlap/heterogeneity.

A second way to look at the timing of these discontinuities is to analyze the thresholds of performance and price that had to be reached before the discontinuities could diffuse. As mentioned in the last paragraph, improvements in magnetic recording density caused users and suppliers to place more emphasis on price than on quality, as shown in Figures 5.1 and 5.2. The easiest way to reduce price and thus cost was to implement the simpler systems, which were also often the smallest. Simply put, the diffusion of these simpler systems required the magnetic recording density of tape to meet certain thresholds of performance and price. Improvements in recording density led to reductions in cost and improvements in performance of both tape and system. Therefore, this chapter focused on minimum thresholds of performance for magnetic recording density. Figure 5.3 defines the performance thresholds needed before a number of discontinuities were successfully commercialized.

Contrasting this analysis with the conventional wisdom, the demand for a new discontinuity such as VCRs had a relatively small impact on magnetic recording density, and thus the theory that demand drives reductions in cost is not very useful for understanding the emergence of discontinuities in magnetic recording and playback equipment. Instead, improvements in recording density, which were driven by previous recording and playback systems, gradually improved the "potential" cost and performance of VCRs, and a growth in demand for VCRs only impacted the cost/performance of magnetic tape as, for example, the tape was modified for a new system. However, once this took place, these improvements probably had a larger impact on the cost of VCRs than did demand for them.

In summary, like Chapter 4, this chapter's analysis can help managers and students analyze the timing and to a lesser extent the diffusion of a potential technological discontinuity. The fact that magnetic recording density had a strong impact on the performance and price of recording and playback equipment, and that it was experiencing exponential improvements in cost and performance, should have convinced firms that this rate of improvement would probably enable discontinuities to emerge. Second, the fact that simple designs were cheaper than more complex designs and that improvements in magnetic recording density enabled their use should have convinced firms that these simple designs had a high chance of success. Third, in terms of recent events, the complicating factor was that improvements in other components also had a large impact on the cost and performance of magnetic or optical-based systems. In addition to the improvements in semiconductor lasers and photodiodes that made CDs and DVDs possible, improvements in other "components" such as hard disks, flash memory, and the Internet made possible listening to music on even more "simple devices" such as MP3 players and mobile phones, and thus caused other discontinuities to emerge.

6

Semiconductors

In this chapter, we look at the history of semiconductors to further explore the time lag between the characterization of the concepts and architectures that form the basis of technological discontinuities and the commercialization of these discontinuities. The chapter makes six related arguments about this time lag.

First, there was a shorter time lag for semiconductors than for computers and magnetic recording and playback equipment because specific discontinuities encouraged the modification of the concepts, which was quickly followed by the emergence of discontinuities based on these modified concepts. Second, as with the discontinuities discussed in Chapters 4 and 5, the time lags for most semiconductor discontinuities involved an insufficient component, which in this case was semiconductor-manufacturing equipment.[1]

Third, unlike the discontinuities in the last two chapters, both the component (manufacturing equipment) and the system (ICs) benefited from geometrical scaling, and thus the contributions from improvements in equipment to the performance and cost of semiconductors were magnified by the large benefits from reducing the scale of transistors, memory cells, and other IC features. Fourth, many of the improvements in "components," in this case manufacturing equipment, occurred relatively independent of improvements in other components. The reason is that, following experimentation with a number of processes and equipment in the 1950s, which involved advances in science, the so-called "planar process" emerged late in that decade; until recently there were few changes in the way the steps in this process and its equipment were organized. Instead, improvements in individual types of equipment, in particular photolithographic equipment,[2] enabled dramatic reductions in feature size and thus led to increases in the number of transistors on an IC.

Fifth, this chapter shows how improvements, particularly those in photolithographic equipment, changed the trade-offs users and suppliers made between price and various dimensions of performance for semiconductors, and how they led to the emergence and diffusion of discontinuities. Users and suppliers considered the trade-offs among many factors when they analyzed new semiconductor and IC designs. The changes in trade-offs are partly explained using the concept of an innovation frontier, where improvements in equipment drive a frontier's movements.

Sixth, we can define a minimum threshold of performance for semiconductor-manufacturing equipment that was needed before technological discontinuities could exceed it and/or fall below a user's maximum threshold of price. This chapter focuses on performance because the basic measure of performance for semiconductors, the number of transistors per chip (i.e., Moore's Law[3]), also considers price. As with the discontinuities discussed in the last two chapters, one caveat here is that finding new applications and implementing new business models for semiconductor discontinuities probably reduced their minimum thresholds of performance.

EARLY SEMICONDUCTORS[4]

The field of semiconductors received little attention from scientists and engineers until John Bardeen, Walter Brattain, and William Shockley demonstrated the first semiconductor transistors in the late 1940s at Bell Labs. Concepts had been earlier proposed for semiconductor transistors, but these were isolated developments. The announcement of the working semiconductor transistor in 1951 and its liberal licensing by Bell Labs caused an explosion in research and development. Companies and universities began investigating different semiconductor materials, transistors made from these materials, and processes for transistor manufacture. Germanium was replaced by silicon, and the so-called planar process emerged in 1959 as the basic method for manufacturing bipolar and later metal oxide semiconductor (MOS) transistors. The planar process depended on scientific advances in semiconductor bands, movements of holes and electrons, and the diffusion of impurities. Although these advances have continued, many argue that until recently most silicon-based transistors and ICs largely depended on the base of scientific knowledge defined by the planar process whereas the organization of transistors and other electronic components into electronic circuits was built on other scientific research.

Looking at these developments in more detail, Table 6.1 summarizes the technological discontinuities in the semiconductor industry and their classification as radical or architectural innovations (through the year 2000). There

TABLE 6.1

Technological discontinuities in the semiconductor industry

Discontinuity	First introduced	Type[a]	Concepts and/or architectures
Germanium bipolar transistor	Early 1950s	Radical: change in material and transistor design (from vacuum tubes)	First prototypes constructed in 1948 and 1955, but planar process (1959) scientifically characterized operation and production of transistors and ICs
Silicon bipolar transistors	Mid-1950s	Radical: change in material	
Bipolar ICs	Early 1960s	Radical: from perspective of electronic circuit	ICs utilized planar process
MOS ICs	Early 1970s	Radical: change in transistor design	Concept first identified as early as 1925, but prototypes not constructed until early 1960s following development of planar process
CMOS ICs	Mid-1970s	Radical: change in transistor design	
Logic ICs	Early 1960s	Radical: from perspective of electronic circuit	Claude Shannon applied Boolean logic to electrical circuits
Memory ICs (new forms of memory came later)	Early 1970s	Radical: new type of circuit element (memory cell)	RAM and ROM (early 1960s); EPROM (1971); EEPROM (1978); flash memory (1980)
Microprocessor ICs	Early 1970s	Radical: from perspective of electronic circuit	CPUs designed for computers in 1940s
ASICs (nonmilitary)	Late 1970s		
ASSPs	Mid-1980s	Radical from perspective of electronic circuit	Standard products (ASSPs) (i.e., constructed from standard components) existed before semiconductor industry; FPGAs (1986)

SOURCE: Tilton (1971), Braun and Macdonald (1982), Malerba (1985), Borrus (1987), Turley (2003), Rowen (2004), Wikipedia (2009a, 2009b), and the author's analysis.

NOTE: MOS, metal oxide semiconductor; CMOS, complementary MOS; EPROM, erasable programmable read-only memory; RAM, random-access memory; ROM, read-only memory; EEPROM, electrically erasable PROM; CPU, central processing unit; ASIC, application-specific IC; ASSP, application-specific standard product; FPGA, field-programmable gate array.

[a]Radical innovations are defined from the perspective of both the transistor and the electronic circuit.

were changes in the choice of materials and in transistor design and organization. From the perspective of a transistor, design and material changes can be defined as radical innovations. From the perspective of an electronic circuit, changes in the organization of transistors can be defined as radical innovations. Table 6.1 classifies transistor-based discontinuities from the perspective of a transistor and electronic circuit–based discontinuities from the perspective an electronic circuit.

As shown in Table 6.1, many of the concepts or architectures that formed the basis of these discontinuities had been identified by the mid-1960s. The first bipolar transistor was demonstrated in the late 1940s at Bell Labs, and the first MOS and CMOS transistors were demonstrated in the early 1960s (the concept was first identified in 1925). The first ICs (logic ICs) grew out of Boolean logic, which had been applied to electrical circuits by the 1930s. Following the use of memory in computers in the 1940s, the first memory cells were demonstrated in the early 1960s, with researchers looking for designs that could be constructed from a smaller number of transistors or other electronic devices such as capacitors. The concept of a microprocessor also came from the computer industry, in this case the use of CPUs in computers.[5] Application-specific ICs (ASICs) such as standard cell designs were based on the concept of building a system from constituent components, much as words are based on letters or printing presses are based on movable type.[6] Application-specific standard products (ASSPs) are merely standard products and thus their concept had been understood for a long time.

The concepts that were identified more recently than 1966 include new forms of memory such as erasable programmable read-only memory (EPROM) in 1971, electrically erasable PROM (EEPROM) in 1978, flash memory in 1980, and new forms of ASICs such as field-programmable gate arrays (FPGAs) in 1986. These built from earlier concepts and thus expanded the knowledge base for semiconductor devices (they are discussed in more detail later). The time lag between the characterization of concepts and architectures and their commercialization was shorter for discontinuities in semiconductors than for computers or magnetic recording and playback equipment.

OVERVIEW OF DIMENSIONS OF PERFORMANCE

Before we can look at what determined the timing of discontinuities, we need to look at the dimensions of performance for ICs and the trade-offs that their users and suppliers made, which can be considered part of the IC technology paradigm. Key dimensions of performance include speed, size, functionality, power consumption, and heat dissipation. Faster speeds enable faster sensing (e.g., of light in a digital camera), processing (e.g., in a microprocessor), storing, and accessing (e.g., in a memory IC) of analog and digital data, which is important in almost every type of electronic product. Functionality can be measured in terms programmability (e.g., in microprocessors) and customizability (e.g., in application-specific ICs). Although power consumption is more important for portable than for nonportable devices, it impacts nonportable devices because high power consumption is related to high heat dissipation and high heat dissipation increases the chance of circuit failure.

Both suppliers and users of semiconductors make trade-offs between price and different dimensions of performance. Power consumption and heat dissipation can be reduced by lower voltages, but these reduce speed. They can also be reduced by placing circuits on separate chips, but this reduces speed and may increase costs. Producers and users also make trade-offs between performance and cost/price, with suppliers and users of low-volume electronic products (e.g., military) often placing more emphasis on performance than on price; the opposite is generally true for suppliers and users of high-volume electronic products. Manufacturing costs are a bigger issue for high-volume electronic products while development costs are a bigger issue for low-volume electronic products.

Semiconductor-manufacturing equipment is used to produce discrete devices and ICs and thus is the most important semiconductor "component." The fabrication of ICs primarily involves the growth or deposition of multiple materials on a so-called silicon wafer using various forms of furnaces, and the formation of patterns in each layer of material using photolithography (in which light only passes through some parts of a "mask") and etching. Although there are different dimensions of performance for each type of equipment, minimum feature size and to a lesser extent defect density typically characterize the performance of a fabrication facility, and photolithographic equipment is usually considered the most important piece of equipment. Minimum feature size typically refers to the minimum gate length in a transistor or the minimum thickness of a layer. Defect density refers to the number of defects per area on a finished chip.

Reductions in defect density and feature size enabled increases in the number of transistors per chip, which is ordinarily referred to as Moore's Law. Reduced defect density enabled a 30-fold increase in chip size (sometimes called die size) between 1970 and 1995, which together with reduced feature size drove increases in the number of transistors per chip. Although other dimensions of performance are also important (see earlier), the number of transistors per chip is often used as the primary measure of performance for semiconductors because more transistors enable improvements in other dimensions of performance, such as increased functionality, reductions in size, and faster speeds through smaller feature sizes. These issues are explored in more detail in the subsequent sections.

DISCRETE TRANSISTORS[7]

Although the emergence of the semiconductor industry in the 1950s depended first and foremost on advances in science associated with the first junction transistor in 1949, processes and equipment borrowed from industries such as aerospace and nuclear energy also supported the commercialization of and rapid improvement in discrete transistors during the 1950s. Pure semiconductor material and thus effective processes for growing semiconductor crystals are essential to semiconductor-based transistor performance, and the creation of the first junction transistor revived interest in these processes. For example, we can say that the revival of Czochralski's crystal-growing approach (first developed in 1917), along with the development of zone refining at Bell Labs in 1950, enabled the purity of germanium to exceed the "minimum thresholds of performance" needed before germanium transistors could exceed the minimum thresholds of performance their first users demanded.

Further improvements in silicon crystal–growing and oxidation equipment led to the replacement of germanium by silicon in transistors. In terms of design trade-offs, the benefits from being able to cover a silicon wafer with a thin layer of oxidation (not possible with germanium) finally exceeded the higher costs associated with the higher melting point of silicon (and thus the higher costs of furnaces), and led to the replacement of germanium with silicon in most semiconductor products, beginning with those for military applications, which still drove spending on semiconductor research.

BIPOLAR ICS[8]

ICs combining multiple transistors and other electronic devices on the same chip were made possible by improvements in equipment used in the planar process. These improvements caused Jack Kilby of Texas Instruments to recognize that the advantages of placing resistors, capacitors, and transistors on the same substrate material (i.e., silicon) would eventually outweigh the advantages of using the optimal material for capacitors (Mylar) and resistors (carbon) in discrete components. It also caused Robert Noyce of Fairchild to recognize that the advantages of using a metal layer to connect multiple transistors on a single IC chip (and not connecting individual transistors with wires) would eventually outweigh the disadvantages of lower yields from placing multiple transistors on a single IC chip.

Subsequent improvements in equipment have reduced feature sizes by more than 100 times and defect densities by more than 1000 times in the last 45 years. The implications of these trends were noted by Gordon Moore in his famous 1965 *Electronics* article that led to "Moore's Law."[9] Almost unchanged from Moore's original figure in that article, Figure 6.1 here shows the evolution in the "optimal" level of integration (see the U-shaped curves) over time. According to Moore, at any given point in time there is an optimal level of integration that minimizes the manufacturing costs per transistor and represents the trade-offs between minimizing the number of ICs in an electronic circuit and maximizing the yield for each IC type. Increasing the number of transistors on a chip reduces the number of chips needed for an electronic circuit, while placing too many on a single IC causes yields to fall.

The curves in Figure 6.1 can be considered innovation frontiers for the early years of the semiconductor industry. Each frontier represents the trade-offs at a given point in time between minimizing the number of ICs and maximizing each IC's yield. Over time, improvements in equipment reduced defect densities and thus increased yields, pushing the frontiers toward the

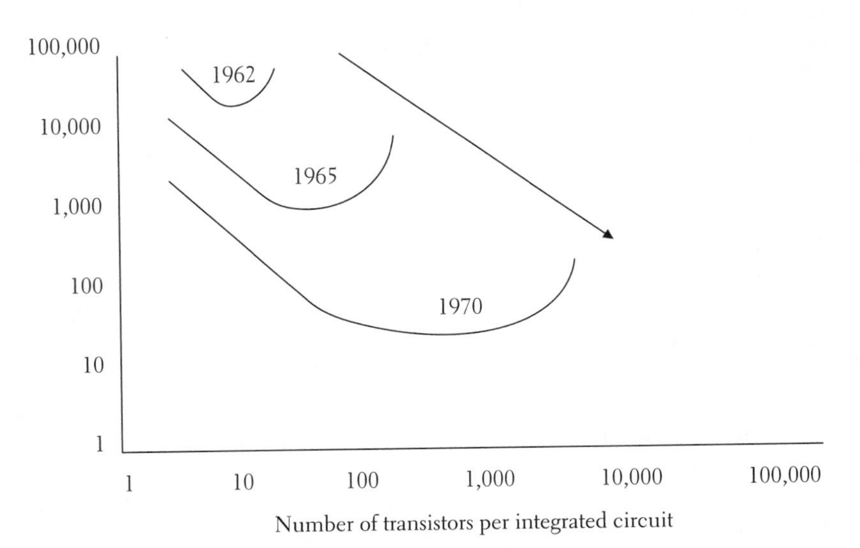

FIGURE 6.1 Innovation frontier for the early years of the semiconductor industry
SOURCE: Adapted from Moore (1965).

 lower right portion of the figure, away from the first IC, which is shown in the upper left. Of course, Figure 6.1 and other representations of Moore's Law simplify these trade-offs since some users (e.g., the military) place more emphasis on performance than on price and thus demand a larger number of transistors than is specified as the optimum. Similarly, some users place more emphasis on power consumption/heat dissipation, programmability, and customizability, and these demands reduce the optimum number of transistors on a single chip. These issues played a role in many of the discontinuities that are to be discussed.

MOS AND CMOS ICS[10]

Metal oxide semiconductor (MOS) and complementary MOS ICs use different transistor designs than those in bipolar ICs. Although the concept that formed the basis for MOSs and CMOSs was first identified in 1925 by Julius Lilienfeld, and the first prototypes were demonstrated in the early 1960s, these ICs were not used in large volumes until pocket calculators and digital watches were commercialized in the early 1970s. This was in part because of their much slower speeds as compared to bipolar logic ICs. However, only MOS and CMOS ICs could provide the low power consumption that pocket calculators and digital watches, respectively, required. In other words, the markets for pocket calculators and digital watches only began to grow when the price of MOS and CMOS ICs fell below the maximum thresholds of price that would enable these devices to fall below the maximum prices that were demanded by consumers. Taking this one step further, the price of MOS and CMOS ICs did not fall below this maximum price threshold until equipment performance had exceeded minimum thresholds (see Figures 6.2 and 6.3).

Further improvements in equipment (e.g., photolithographic equipment) used in the planar process and their resulting impact on feature sizes (see Figure 6.2), as well as improvements in the number of transistors per chip (see Figure 6.3), which increased the importance of power consumption, eventually favored CMOS over MOS ICs, and both of these over bipolar ICs, in most applications. In particular, the increasing number of transistors on a chip caused power consumption to become a major problem in most electronic products and thus favored the lower power consumption of CMOS over MOS and bipolar ICs. CMOS transistors were first used in DRAMs (dynamic random-access memories) in 1986 in a 1MB device, and their use in IC production rose from 40 percent in 1988 to 80 percent in 1994.[11]

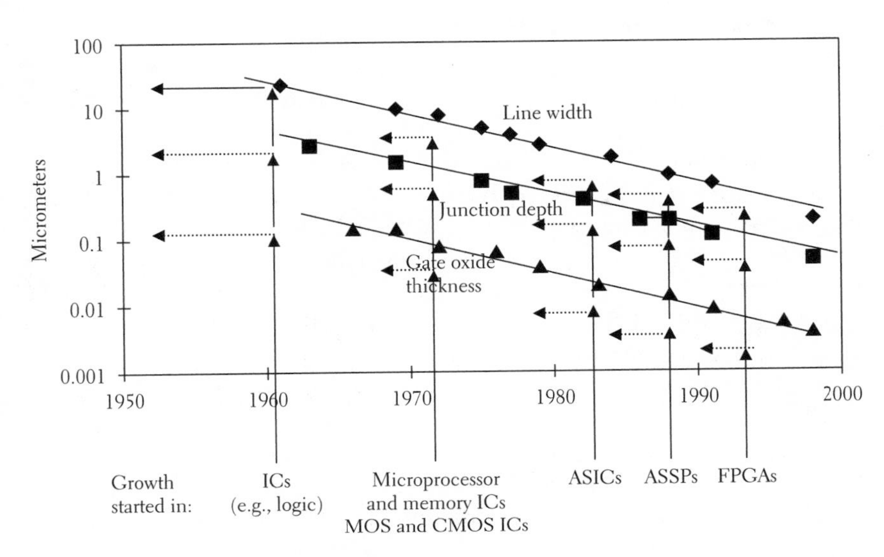

FIGURE 6.2 Minimum thresholds of component/equipment performance for IC discontinuities

SOURCE: O'Neil (2003) and author's analysis.

NOTE: MOS, metal-oxide-semiconductor; CMOS, complementary MOS; ASIC, application-specific IC; ASSP, application-specific standard product; FPGA, field-programmable gate array.

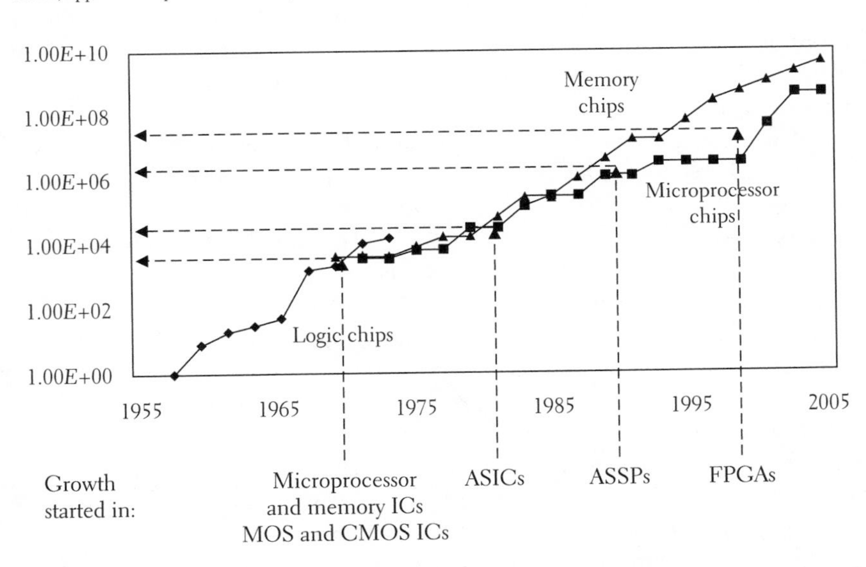

FIGURE 6.3 Minimum thresholds of equipment performance in terms of number of transistors per chip

SOURCE: International Technology Roadmap for Semiconductors (ITRS; various years), ICKnowledge (2009), and author's analysis.

MICROPROCESSORS AND MEMORY[12]

A microprocessor incorporates most of a computer's central processing unit (CPU) on a single IC; a memory IC stores data, including a microprocessor's program. As improvements in equipment were driving changes in the choice of transistor design (between bipolar, MOS, and CMOS) in the 1970s, they were also improving the economic feasibility of microprocessor and memory ICs. As discussed in Chapter 4, few changes have been made to the basic architecture of computers where microprocessors are just a scaled-down version of a CPU. For memory chips, one way to increase the number of memory bits they hold is to design a memory cell that requires fewer transistors and other devices; however, reductions in the number of transistors per memory cell have been at the single-digit level while increases in the number of transistors per chip have been in orders of magnitude. Therefore, we can say that microprocessor and memory ICs were introduced when a minimum threshold of performance for equipment was reached that technically enabled a single IC chip to hold a computer's CPU or a specific number of memory bits (the first memory chip had about 1000 memory cells).

The fact that microprocessors were first used in products that were not driving demand for bipolar, MOS, or CMOS ICs probably reduced the minimum thresholds of performance for microprocessors and the equipment used to produce them. Although the first orders for a microprocessor were placed by Japanese calculator manufacturers, the rapidly growing market for calculators enabled special-purpose ICs to be used in them. Instead, a large number of low- to mid-volume applications such as aviation and medical and test equipment were the initial markets for microprocessors.

Trade-offs and innovation frontiers can also explain the diffusion of microprocessors and, to some extent, memory chips. Semiconductor users made trade-offs between development costs, speed, and manufacturing costs when they compared logic chips, microprocessors, and custom ICs in the 1970s. Custom ICs had the highest speeds and lowest manufacturing costs but also the highest development costs. Logic chips had the lowest development costs but also the lowest speeds and highest manufacturing costs. Microprocessors provided an intermediate solution between custom and logic ICs. Continued increases in the number of transistors per chip increased the advantages of microprocessors over custom chips (lower development costs) and logic chips (better performance), as discussed in more detail in the next section. Furthermore, new forms of memory and programming tools facilitated the storage and accessing of programs and included ROMs, PROMs, EPROMs, EEPROMs, and flash memory, which were made possible by increases in transistors per chip and by the characterization of new concepts(to be discussed).

Some of these basic design trade-offs are shown in Figure 6.4, a modified form of Figure 6.1 that portrays the evolution in the innovation frontier for the semiconductor industry after 1970. At any given time, the maximum number of transistors per chip is a function of the chip's functionality/complexity. More transistors can be placed on DRAMs than on microprocessors (and on ASICs and FPGAs, which will be addressed) because DRAMS have less functionality than do microprocessors. Over time, improvements in semiconductor-manufacturing equipment reduced feature sizes and defect densities, and pushed the innovation frontier toward the upper right portion of Figure 6.4. Although such data is not shown in this chapter's figures, innovation frontiers can also be drawn for the different types of memory just mentioned and for the trade-offs between, for example, the number of transistors per chip and the die size for a single form of memory such as DRAM.

Nevertheless, some of the new forms of memory depended on the development of new concepts, motivated by the success of earlier forms of memory such as DRAM and SRAM (static RAM). For example, Intel's Dov Frohman developed the first EPROM in 1971 while investigating failure rates of DRAMs. His development caused Intel and other firms to introduce EPROMs and further investigate the science behind them. This research led to Intel's George Perlegos' development, in 1978, of a method for erasing programs with electrical signals as opposed to ultraviolet light, which formed the basis for

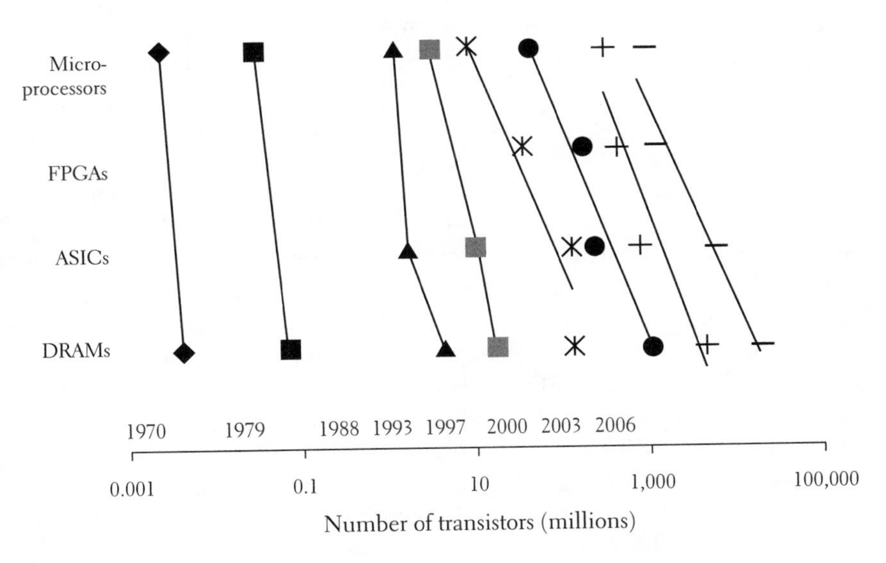

FIGURE 6.4 Innovation frontier for the semiconductor industry (after 1970)

EEPROMs. Also driven by a desire to facilitate the use of microprocessors in new (including portable) applications, Toshiba's Fujio Matsuoka developed the concept that formed the basis for flash memory in 1980, which began to experience growth in the early 1990s with the diffusion of laptop computers.[13]

APPLICATION-SPECIFIC ICS[14]

Application-specific ICs (ASICs) are semi-custom chips that have largely replaced logic chips. They provide an intermediate solution between microprocessors and full-custom chips. Although ASICs have lower functionality in terms of programmability than do microprocessors, they are faster and have higher densities (i.e., the number of transistors per chip) but are slower and have lower densities than full custom chips. Furthermore, each ASIC type, whether a standard cell, gate array, or FPGA, provides different levels of customization and thus different speeds, densities, and prototype times and costs. Basically, as we move from standard cells to gate arrays and FPGAs, prototype cost rises but so does performance, while unit costs fall as long as volumes rise.[15]

The concepts of minimum thresholds of equipment performance and trade-offs can be used to explain ASICs, as with microprocessors. First, the increasing number of on-chip transistors enabled ASICs to replace logic chips in many applications because more transistors made it difficult to design logic chips that used the full extent of integration possible from Moore's Law but could still be considered "general-purpose." Second, more transistors per chip caused development costs of ICs to rise and rise faster than improvements in computer-aided design tools.[16] In both cases, we can think of the minimum number of transistors (see Figure 6.3) at which a logic chip is no longer considered general purpose and instead is considered custom, and at which the higher development costs of custom chips begin to favor ASICs.

Some of these basic design trade-offs are illustrated in the innovation frontiers in Figure 6.4. Although there are analyses that consider FPGAs to be a form of ASIC, the figure differentiates them from ASICs such as gate arrays and standard cell designs because they have greater functionality. It shows how the differences in maximum transistors per chip between different IC types have grown over the last 40 years and thus expanded the niches for gate arrays, standard cells, and FPGAs. For example, the ratio of the maximum number of transistors for DRAMs to those for microprocessors increased from less than 2 in 1970 to a high of 25 in 2000 and to 22 in 2008.

FPGAs had a shorter time lag between concept and commercialization than many of the discontinuities covered in this chapter. The concept of an FPGA was not fully articulated until the mid-1980s[17] and then was quickly implemented. It was driven by a desire to reduce the development costs of ASICs, but grew from ROM, which was first introduced in 1970 by Texas Instruments, and evolved through PROMs and programmable logic devices (PLDs) throughout the decade. More transistors per chip made the increased functionality of the FPGA, when compared to the PLD, possible.

APPLICATION-SPECIFIC STANDARD PRODUCTS[18]

The term *application-specific standard product* (ASSP) refers to standard IC chips designed for a specific system/product and often for a specific standard module in that system/product. ASSPs' designs are somewhat similar to those of microprocessors or ASICs since many of them are microprocessors configured for a specific application, often using ASIC design techniques. The difference between ASSPs and ASICs is that the former are standard ICs used by different producers of a final electronic product.

The primary driver of growth in the ASSP market is the increasing volumes of electronic products. Until the market for personal computers began to grow in the 1980s, electronic products that were both complex and produced in high volumes did not exist. Digital calculators and watches were produced in high volumes in the 1970s, but by the 1980s they did not require the full number of transistors that could be placed on a single IC. On the other hand, complex electronic products such as telecommunication switches and mainframe computers were produced in low volumes and so were assembled from a combination of ASICs and standard logic chips, microprocessors, and memory.

The increasing volumes of electronic products are a result of Moore's Law and thus of improvements in equipment used in the planar process. Although at any given point in time, increases in volume likely lead to lower costs, electronic products cheap enough to be in high demand did not emerge until IC prices had fallen and their performance had risen to some specific level (see Figure 6.3). In other words, IC prices had to fall below a maximum threshold before prices of the final electronic products could fall below a maximum threshold. This is another example of the fact that cumulative production does not explain cost reductions. It is not the volumes of electronic products that drive cost reductions in semiconductors and ICs; rather, it is improvements in semiconductor-manufacturing equipment and the benefits from reduced scale, which thus enable larger numbers of people to purchase electronic products such as PCs, mobile phones, and digital set-top boxes.

Consider PCs and the ASSPs that are used in them. Intel gradually customized its PC microprocessors as the market for PCs grew in the 1980s. Thus its microprocessor can be defined as an ASSP for PCs. Other firms provided basic chip sets for PCs that were compatible with Intel's microprocessors, and that enabled assemblers to offer so-called IBM clones in the mid-1980s. As the falling prices for memory chips enabled bitmap displays, firms offered special processors in the 1990s that handled this data and were also compatible with Intel's microprocessors.[19] At each point in the growth of the PC market over the last 30 years, one can estimate when the number of transistors on a chip had reached the minimum threshold of performance for the price of these ICs to fall below a maximum threshold, and thus for the price of a PC to fall below its maximum threshold for a new group of users.

THE FUTURE

Further reductions in feature size will enable further increases in the number of transistors per chip and these increases will lead to new organizations of transistors such as systems on a chip and configurable processors. On the other hand, we are probably approaching the limits to Moore's Law and thus the limits to further reductions in feature sizes with the current technology paradigm. Questions about these limits began to emerge in the 1990s as feature sizes fell below the wavelength of visible light, requiring smaller-length light sources such as ultraviolet.[20] At some point in time, new and more expensive methods of forming patterns on wafers such as e-beam lithography will be required. E-beam lithography is already used to make the photomasks for photolithography, and its equipment is much more expensive than photolithographic equipment. Other problems exist with the conductivity and capacitance of interconnect, and solutions include a change from aluminum to copper for interconnect and the introduction of new materials for insulators.[21] Larger and more fundamental problems include quantum mechanical effects. These problems suggest that a new paradigm must be found, and some optimists believe that promising replacements include memristors, phase change memory, 3D transistors or memory cells, organic transistors, molecular transistors, and quantum dots.

On the more positive side, improvements in semiconductor-manufacturing equipment are and will likely continue opening up completely new markets that may become larger than the existing ones. These include bioelectronics, microelectronic mechanical systems (MEMSs), and tissue engineering. Bioelectronic ICs can sense and analyze biological material in, for example, point-of-care diagnostics and machines for sequencing and synthesizing DNA for drug discovery. This will provide better drug delivery in IC-controlled smart pills and better control of artificial implants. Also, reducing the scale of devices such as microfluidic channels on bioelectronic ICs has contributed to multiple orders-of-magnitude reductions in the cost of sequencing and synthesizing DNA.[22] Similarly, MEMS, which are produced using some of the same technology used to construct ICs, are used in motion sensors for Nintendo's Wii, in nozzles for ink-jet printers, in microgas analyzers, and in the building blocks for optical computing (e.g., waveguides, couplers, resonators, and splitters). Reductions in scale increase the sensitivity of these devices and the number of transistors available for processing information. For tissue engineering, the scaffolding that provides a structure for cell growth in a new organ is primarily produced using processes and manufacturing equipment borrowed from the semiconductor industry. Reductions in feature sizes enable new organs to better fit the tissue contours of a host.[23]

Because the feature sizes for these technologies have reached the nanometer level, the equipment and processes used to produce them are now referred to as *nanotechnology*. The term applies to any material that involves features smaller than 100 nanometers (the minimum width or spacing for a "line" in Intel's latest microprocessors is about 32 nanometers), and such materials include very small pores for membranes or nanoparticles both of which have surface area–to–volume ratios that are very large. In combination with the notion of self-assembly, which is the key to realizing the full benefits of nanotechnology, the ability to process even smaller features, such as at the molecular or atomic level, may enable the economical manufacture of carbon nanotubes, graphene, and other materials that utilize molecular and atomic forces and thus, because of scaling, have high levels of performance on a per-weight basis. The benefits to the semiconductor industry from previous reductions in scale may be nothing compared to the benefits from further reductions, and the notion of scaling can help us better understand the opportunities for nanotechnology.

CONCLUSIONS

The chapter used the history of semiconductors to further explore the time lag between the identification of the concepts and architectures that form the basis of technological discontinuities and the commercialization of these discontinuities. As previously noted, complementary technologies that sometimes depend on novel combinations of components are usually the explanation. The chapter showed that in the semiconductor industry the time lag for many discontinuities resulted from a lack of one type of component, semiconductor-manufacturing, particularly photolithographic, equipment. After the planar process defined the basic process and the relationship between different types of equipment, improvements in photolithographic equipment (more than novel combinations) enabled reductions in feature size (and in density of defects) as well as increases in the number of transistors per chip and the commercialization of different ICs such as logic, memory, microprocessors, ASICs, and ASSPs. This occurred long after the concepts that formed the basis of these discontinuities had been characterized. Furthermore, many of these discontinuities, such as memory and microprocessor ICs and ASICs, can be defined as new industries.

This does not mean that research on semiconductors or novel combinations of equipment did not affect competition between firms. The management literature on this industry[24] describes how firms such as Intel, Texas Instruments, NEC, Toshiba, Samsung, and others conducted research, found new customers, and made good and bad design decisions, as well as how these decisions impacted the performance and price of products and thus industry competition. Furthermore, the success of specific products defined as technological discontinuities or the implementation of process improvements also depended on design decisions that might be defined as novel combinations of "components." The point here is that these design decisions had a much smaller impact on the timing of discontinuities than did reductions in feature size and defect density, which were primarily driven by improvements in semiconductor-manufacturing equipment. Improvements of almost ten orders of magnitude in the number of transistors per chip were achieved from reductions of four to five orders of magnitude in feature size.

As in the previous two chapters, one way to analyze the timing of system-based discontinuities is to look at the trade-offs that suppliers and users were making between price and performance and between various dimensions of performance for a system, and to look at the impact on these trade-offs of improvements in components. For example, because of reductions in the defect density of transistors, the benefits of integration exceeded its extra costs (use of nonoptimal materials and reductions in yields), and this led to the emergence of the IC as a discontinuity. Fewer improvements in defect density would have delayed or perhaps have prevented the emergence of the IC.

For subsequent ICs, users and suppliers made trade-offs between price, speed, power consumption, and programmability. Moreover, these trade-offs were influenced by user development costs and improvements in semiconductor-manufacturing equipment. For example, increases in the number of transistors per chip changed the trade-offs that both users and designers made with respect to power consumption, heat dissipation, and speed, and this led to the discontinuities of MOS and CMOS transistors. Also, further increases in the number of transistors per chip changed the trade-offs that users and designers made with respect to development costs and functionality.

They elevated the importance of development costs and reduced the importance of functionality (which impacts speed); thus they changed the costs/functionality trade-offs that designers were making. This led to greater use of microprocessors and ASICs. More gradual increases in the number of transistors per chip would have slowed the transition from bipolar to MOS and CMOS transistors and the emergence of microprocessors and ASICs. These examples highlight another advantage of using the term *trade-offs* instead of Christensen's *technology overshoot*, which would have made it more difficult to characterize any of the preceding examples.

Also as in the previous two chapters, one can define a minimum threshold of performance for semiconductor discontinuities and for the photolithographic equipment that made them possible. This was done for the number of transistors per chip and feature sizes, both of which reflect the performance of semiconductor-manufacturing equipment. Figures 6.2 and 6.3 define the minimum thresholds of equipment performance needed before the first bipolar ICs, MOS and CMOS ICs, microprocessors, memories, ASICs, and ASSPs became economically feasible. One of the challenges of such estimations, particularly ex ante, is that typically not all of the applications are known, and this was certainly the case with discontinuities in the semiconductor industry. MOS and CMOS ICs and microprocessors were first used in new applications and often by new users. Finding these new users and their applications probably made it possible for the discontinuities to diffuse before their performance (e.g., speed) had reached the performance of existing ICs.

One large difference between this chapter and the previous two chapters is that some semiconductor discontinuities—for example, EPROMs, EEPROMs, and flash memory—exhibited a much shorter time lag than did the discontinuities for computers and magnetic recording and playback equipment. This may be because the large demand for microprocessors and their associated memory was unexpected and thus it motivated specific types of scientific research. Rosenberg argues that something similar occurred in the 1950s following the development of the junction transistor at Bell Labs in the late 1940s. The unexpected demand for transistors that followed caused many to realize that solid-state semiconductor physics was an important area for research, and both personnel and funds were quickly found to support it.[25] A similar phenomenon seems to have occurred following the unexpected growth in the market for microprocessors.

Another large difference between semiconductor discontinuities and the discontinuities in computers and recording equipment is that the newer discontinuities in semiconductors have not completely displaced earlier ones; instead, many of them co-exist. For example, mainframes and, to a greater extent, minicomputers were largely replaced by PCs; reel-to-reel and 8-track players were largely replaced by cassette and later CD players; and helical-scan players were replaced by DVD players. On the other hand, microprocessor and memory ICs, ASICs, and ASSPs continue to exist side by side. One explanation might be that the network effects associated with computers and audio and video recording and playback equipment were stronger than those associated with semiconductors and led to more winner-take-all situations. A second explanation might be that there is a larger preference overlap between different users of computers or magnetic tape storage systems than between different users of semiconductors. Figures 4.2, 5.2, 5.3, and 6.4 partly support this explanation. Figures 4.2, 5.2, and 5.3 show how the trade-offs between the major dimensions of performance and price decreased over time, but this has not been the case in semiconductors. Instead, the trade-offs between functionality and the number of transistors in Figure 6.4 have actually increased in that the differences in the number of transistors per chip between low (DRAM) and high (microprocessor) functionality have increased. These increasing differences suggest that different users are making very different trade-offs between these two dimensions of performance and that these different users enable the co-existence of different types of ICs.

III
Opportunities and Challenges for Firms and Governments

Technological discontinuities and new industries provide users, firms, and governments with opportunities and challenges. While the chapters in Parts I and II focused on when a discontinuity might become economically feasible, and thus implied that users easily adopt and new firms easily introduce new technologies, Chapters 7 and 8 summarize the complexities of industry formation and thus the challenges. These challenges may delay the adoption and diffusion of new technologies, or they may enable new entrants or even new countries to dominate a new industry. Chapter 7 focuses on what determines the number of firms that can co-exist in an industry; Chapter 8 explores the challenges that different industries present to firms, governments, and new industry formation in general. Chapter 7 also introduces another major way in which new industries are formed, vertical disintegration. Vertical disintegration emerges through competition among firms, it enables late entry, and it often allows a new set of firms to dominate activities that are based on a new concept and that represent significant sales. In this way, these activities can be defined as a new industry.

7

Competition in New Industries

Technological discontinuities and new industries provide existing firms and new entrants such as entrepreneurial start-ups with profit-making opportunities and new challenges. The reason for the frequent success of entrepreneurial start-ups is that technological discontinuities often destroy an existing firm's capabilities.[1] These capabilities can be in many areas, including R&D, manufacturing, marketing, and sales, and their destruction may be associated with the emergence of new customers. For example, Christensen argues that incumbents often fail when a low-end innovation displaces the dominant technology (thus becoming a disruptive innovation) largely because the low-end innovation initially involves new customers and serving them requires new capabilities.[2] Analysis of the timing of technological discontinuities, the subject of Part II, can help firms identify and prepare for them through, for example, identifying appropriate customers and creating the relevant new capabilities to serve them.

Some research has found that the total number of firms in an industry often declines quickly, in a so-called shakeout, which can occur even when the total market for a product continues to grow. For example, there were large shakeouts in the U.S. automobile, television and personal computer industries in the 1910s, 1950s, and 1980s, respectively, even though the markets for these products continued to grow long afterward.[3] Similar shakeouts have occurred with microprocessors and memory ICs, ASICs, videocassette recorders, CD players, MP3 players, minicomputers, LCDs, and mobile phones, and they will probably occur with tablet computers and smart phones.[4]

Understanding when and why the number of firms quickly declines is essential to understanding the challenges and opportunities for firms in new industries. Since the number of firms in an industry is one measure of opportunities, those that enter the industry when that number is increasing probably have a greater chance of success than those that enter when the number is falling. This chapter looks at when and why shakeouts occur, when entry is sometimes possible after a shakeout, and thus how the opportunities in a new industry change over time. It does this by first examining Klepper's theory of shakeouts and then applying it to the information technology (IT) and semiconductor sectors.

WHAT DRIVES A SHAKEOUT?

Although it used to be thought that the emergence of a "dominant design" triggered a shakeout because it defined a set of necessary capabilities that firms must have to effectively compete,[5] most now agree that Klepper's theory of economies of scale[6] provides a better explanation. Klepper showed that economies of scale in R&D cause small firms to exit or be acquired. Economies of scale in R&D (or in other activities) favor firms with large sales because large firms can spend more on total R&D than smaller ones can. Thus, those with initially higher R&D spending are able to introduce more and better products and so enjoy more sales. Higher sales enable a firm to spend more on R&D and introduce more products, leading to yet more sales. This positive feedback between R&D spending, the number of products, and the amount of sales causes large firms to grow faster than small ones and small ones to be acquired or to exit the industry.

Of course, creating this positive feedback is not just about quickly introducing new products and increasing a firm's R&D budget. The products must sell, and they must sell for a profitable price to enable a firm to fund more R&D. Introducing successful products requires that firms effectively manage product development, part of which involves good timing. If a firm introduces a product too early, before it can offer a superior value proposition, sales or profits will not appear and the firm will not benefit from R&D economies of scale. Furthermore, a firm must offer the right design for the right customer, which is complicated by the emergence of new market segments over time. As discussed in Parts I and II, we can analyze when a new technology might offer a superior value proposition to an increasing number of users by analyzing the rates of improvements in its system and in that system's components.

On the other hand, late entry is sometimes possible. First, the existence of submarkets can reduce R&D economies of scale because each submarket requires different types of R&D, and this can prevent a shakeout. For example, Klepper concluded that large number of submarkets for business jets and lasers prevented a shakeout in these industries. Second, even after a shakeout has occurred, entry is still possible as vertical disintegration emerges. Klepper concluded that the emergence of independent equipment suppliers in the early-to mid-20th century reduced the cost of R&D for the manufacturers of final products; this led to a second wave of firms becoming manufacturers of products such as petrochemicals, zippers, and diapers. Furthermore, this vertical disintegration defined new industries in which a new set of firms performed a new set of activities that eventually engendered a significant amount of sales.

The following two sections show how submarkets enabled a large number of firms to co-exist in the U.S. IT and semiconductor sectors, and how vertical disintegration enabled the late entry of many firms, including de novo entrepreneurial start-ups—that is, those founded specifically to enter a new industry. Evidence for the growing number of opportunities in the IT sector is the more than 100,000 firms in it in 2005, up from an estimated 50,000 worldwide in 1993 and 2,500 in 1965.[7] While the numbers are smaller for the semiconductor sector, analyses suggest that more than 1000 U.S. firms[8] have been founded to sell semiconductors, which is far more than ever entered the U.S. automobile, television, or personal computer industries.

The following sections analyze a broader range of vertical disintegration than just manufacturing equipment. This is because, while independent equipment suppliers were a rare example of vertical disintegration in the early 20th century,[9] several types of vertical disintegration emerged in the second half of the century in many sectors, including IT, semiconductors, broadcasting, consumer electronics, mortgage banking (e.g., securitization), construction, and pharmaceuticals (e.g., biotechnology).[10] Many of these can be defined as new industries. Building from work by Jacobides, his colleagues, and others, we can say that technological, institutional, regulatory, and social changes influence how economic agents divide up work and thus impact vertical disintegration. For example, pressures for increased profits or reduced risks, recognition that certain activities fall outside a firm's main capabilities, and falling costs of contracts have encouraged firms to outsource services, such as janitorial, credit collection, and training, that are now considered large industries in themselves.

For high-technology products, vertical disintegration often depends on the transaction costs and capabilities associated with different firms developing different parts of a complex system. The reason is that reductions in transaction costs can reduce both the costs of having work done by multiple agents and the importance of integrative capabilities.[11] These reductions can come from open modular designs, open standards, and legal, regulatory, and firm decisions,[12] in either a top-down or a bottom-up process.[13] In modular design, the interfaces that determine how the functional components or "modules" in a product or process will interact are specified to enable the substitution of components. Design rules specify these interfaces.[14] The term *standard* or *interface standard* is often used to define the way in which these different modules interact, particularly when products from different firms are compatible with the same interfaces and thus use the same design rules.[15] One reason for the importance of modular design and vertical disintegration is that they reduce development costs by enabling a better breakdown in work and by enabling different modules to be shared by different products or even, in the case of vertical disintegration, different firms.

THE IT SECTOR

The emergence of vertical disintegration in the IT sector basically followed the evolution of computers that was described in Chapter 4. Following the appearance of a discontinuity such as the mainframe, the minicomputer, the PC, and the workstation, vertical disintegration enabled the late entry of start-ups that focused on software, peripherals, remote services, and even computers themselves. The main differences between this chapter and Chapter 4 are that we focus on pre-1995 activities here and use the term *client-server systems* to represent a combination of workstations and PCs within a single firm. Such combinations later evolved into a relationship between computers and servers in the overall Internet.

Table 7.1 summarizes the relevant interface standards for each of the four technological discontinuities. Open standards emerged for the interfaces between computers and peripherals, between operating systems (OSs) and application software, and between computers and some services (e.g., remote services), but they did not emerge for the interface between the OS and CPU (central processing unit). Thus the former facilitated the entry of new firms while the latter did not. For example, the release of IBM's System/360 in 1964 and IBM's announcement of software unbundling in 1969 (in response to government pressure, which was not applied to Microsoft in the mid-1990s) led to open standards and thus enabled the late entry of peripheral, application software and a few mainframe computer start-ups. Similarly, the release of the 16-bit minicomputer in 1970 increased the compatibility both within minicomputers and between minicomputers and mainframes, and thus it facilitated the entry of application software developers, value-added resellers, and even minicomputer manufacturers.[16]

TABLE 7.1

Standards for discontinuities/systems in the IT sector (1950–1995)

System	Interface between	Standard or product representing standard	Year released	Open versus closed
Mainframe computer	Computer and peripherals	IBM System/360	1964	Open
	OS and application software	IBM's unbundling of software and hardware	1969	Open
	OS and CPU	IBM System/360	1964	Closed
	Computer and remote services	Various, enacted after Hush-A-Phone case	1956	Open
Minicomputer	Computer and peripherals	16-bit word length (multiple of 8 bits)	1970	Open
	OS and application software			
	Computer and value-added reseller			
	OS and CPU	None	NA	Closed
PC	Computer and peripherals	IBM PC (later called Wintel standard)	1981	Open
	OS and application software			Initially open
	OS and CPU			Closed
Client-server system	Computers and Internet	TCP/IP	1980	Open
	Computers and LAN	Ethernet	1980	Open
	Computers and users	Windows	1990	Closed
	Application software from different systems/discontinuities	None	NA	Closed
	Workstation OS	UNIX	1970	Open

SOURCE: Flamm (1988), Rifkin and Harrar (1988), Langlois (1992), Steffens (1994), Pugh and Aspray (1996), Ceruzzi (1998), von Burg (2001), and Campbell-Kelly (2003).

NOTE: CPU, central processing unit; OS, operating system; LAN, local area network; NA, not applicable.

However, while there was a shakeout of mainframe and minicomputer manufacturers, there was none with software suppliers because the large number of submarkets enabled many of them to co-exist. According to Campbell-Kelly's book on the software industry,[17] 2928 software packages were listed in a software catalogue in 1974, many of which were supplied by different firms. Forty-two percent were industry-specific and the rest were cross-industry offerings, with both reflecting the large number of submarkets

Open standards for remote services came as a result of a D.C. Circuit Court of Appeals decision (238 F.2d 266 (D.C. Cir. 1956)) for the manufacturer of the Hush-A-Phone in 1956.This decision enabled firms to access remote services through AT&T's phone lines, modems, and "dumb" terminals.[18] Some firms provided time-sharing services in which corporate customers paid for access to mainframes and later minicomputers. Others provided payroll, accounting, credit card, sales, financial, health care, and other processing services for corporate customers,[19] which reflected a large number of submarkets for information. Still other firms provided the dumb terminals were used by corporate customers to access time-sharing and processing services. The impact of these remote services and dumb terminals on entrepreneurial opportunities is reflected in the large number of successful firms that provided them: according to one source,[20] between 30 and 40 percent of the top 50 U.S. IT firms in terms of revenue in the late 1970s.

The release of the IBM PC in 1981 defined relatively open standards for the interfaces between microprocessor, OS, peripherals, and application software. This enabled PC clone manufacturers such as Dell and Gateway to use the same externally available modules that IBM used. These modules included Microsoft's OS, Intel's microprocessor, and application software and peripherals from many firms. However, the number of firms founded after the release of the PC was not as high as that seen for previous discontinuities.[21] A likely reason is the ultimate lack of openness in the standards for interfaces between microprocessor, OS, and application software. As just mentioned, these interfaces were relatively open when IBM first released its PC in 1981, but observers argue that Microsoft (and to some extent Intel) gradually increased its control over them in the late 1980s and early 1990s, reducing the entrepreneurial opportunities for other providers. For example, initially successful software from WordPerfect (word processing), Lotus (spreadsheet), and Borland (database) was largely replaced by software from Microsoft in the 1990s, and only two PC application software providers, Adobe and Novell, remained in the top 50 software firms by 2002.[22]

For client-server systems, open standards emerged in workstations (UNIX), the Internet (TCP/IP), and local area networks, or LANs (Ethernet).This was before Microsoft's release of Window 3.0, which cannot be considered an open standard, marked the official beginning of client-server computing in the early 1990s. Unlike other IT discontinuities, these standards emerged through a bottom-up process in which no single set of designers controlled the definition of client-server architecture. Instead, each standard was developed in an independent and fairly decentralized manner that enabled entry of vertically disintegrated suppliers of LAN/routers, workstations, and system software.[23] By 1995, almost 40 percent of the top 50 U.S. IT firms in terms of revenue were providers of client-server products and software.[24]

In summary, vertical disintegration and submarkets enabled a large number of firms to enter and co-exist in the U.S. IT sector because they reduced the economies of scale in R&D and other activities necessary for firms to survive. This enabled large numbers of de novo entrepreneurial start-ups to enter and succeed in the IT sector as well. For example, according to one source, the number of de novo entrepreneurial start-ups among the top 50 IT firms grew from 34 percent in 1980 to 60 percent in 1995, and only 10 firms (which were all de alios) from the top 50 in 1975 still remained there in 1995.[25] On the other hand, a combination of economies of scale in R&D and a smaller number of submarkets reduced the number of opportunities for manufacturers of OSs, microprocessors, video processors, and other standard ICs. The growth of the Internet and the World Wide Web in the mid-1990s probably strengthened these trends. An increasing number of content providers and niche suppliers of software can co-exist in the global Internet, where a few "vertically disintegrated layers," such as search engines, operating systems, and some utility/cloud computing services, are dominated by a small number of large and very profitable firms.

Vertical disintegration has also enabled the emergence of many new industries that consist of de novo entrepreneurial start-ups. For example, hard disks, optical discs, printers, monitors, and perhaps even keyboards can be considered industries because these products are based on a new concept, they are supplied by a different set of firms, and many of them now represent many billions of dollars in sales. Also, many have been affected by technological discontinuities that may have further defined new industries, such as smaller disk drives and laser printers.

THE SEMICONDUCTOR SECTOR

As in the IT sector, there has been continued growth in the number of firms in the U.S. semiconductor sector, which probably now number in the thousands. The fact that semiconductors represented the fourth largest recipient of U.S. venture capital funding in 2005 (following IT software, biopharmaceuticals, and telecommunication) suggests that many entrepreneurial opportunities are still being created. Also as in the IT sector, vertical disintegration and a large number of submarkets have supported this increase in firms while shakeouts have occurred in many of the vertically disintegrated layers.[26] The large number of submarkets in the semiconductor sector can be seen in the many electronic systems for which different types of ICs, particularly application-specific standard products (ASSPs), are being developed and offered. However, since the types of data available for semiconductors are different from those for IT, analyses of the semiconductor sector have focused on the relative ability of de novo and de alio firms to exploit the entrepreneurial opportunities that vertical disintegration has provided.

The semiconductor sector was initially dominated by de alio firms that entered it from an electronic systems sector such as the military, broadcasting, or telecommunication. Although de novo firms represented 31.3 percent of early semiconductor suppliers, their representation among the largest suppliers did not begin to increase until much later (see Table 7.2), mainly because semiconductor suppliers initially needed capabilities in both semiconductor and electronic system design and in both semiconductor design and manufacturing. System design capabilities were needed to define the types and combinations of transistors, while manufacturing capabilities were needed because standard design rules and thus contract manufacturing did not yet exist.[27]

Two forms of vertical disintegration enabled de novo firms to grow and join the largest U.S. semiconductor suppliers (see Table 7.2). First, the gradual emergence over the last 40 years of so-called "standard" semiconductor modules such as logic, memory, microprocessors, application-specific integrated circuits (ASICs), and ASSPs,[28] which were defined as discontinuities in Chapter 6, enabled system suppliers to design their electronic systems as modular partly as a way to reduce system development costs. These modules include open design rules or open interface standards that facilitate incorporation into electronic systems, often in combination with other standard modules. For ASICs, which are semi-custom products, standard transistor arrays, predesigned circuitry, and design tools were the "standard modules," which electronic system firms customized. For ASSPs, there were different standard modules and different standard interfaces for different electronic systems. In summary, standard modules reduced the transaction costs of different firms' designing semiconductors and electronic systems and lessened the importance of electronic system design for semiconductor suppliers. Thus vertical disintegration between electronic systems and semiconductors appeared.

Second, reductions in transaction costs associated with design and manufacturing reduced the importance of design-to-manufacturing integration,

TABLE 7.2

Percentage of de novo semiconductor firms

Databases/lists	1955	1965	1983/1984	1995	2005
Most suppliers	31.3	31.3	31.3	82	—
Top 12	8.3	16.6	50	58	58
Top 35	—	—	51	69	69

SOURCE: 1955 and 1965 data are based on Tilton (1971); 1983/1984 data are based on United Nations (1986) and Dataquest (Yoffie 1993); 1995 data are based on Integrated Circuit Engineering (ICE; 1996); 2005 data are based on Electronic Business (Edwards 2006) and Compustat.

and thus enabled vertical disintegration between so-called design houses and foundries. These reductions were driven by what is known as "dimensionless and scalable design rules," and by standard CAD tools and the Internet. Because design rules define geometrical relationships between line widths, material thicknesses, power consumption, and speed, they reduced the need for communication between design and manufacturing, even though designs had to be updated for smaller feature sizes. The rising cost of new fabrication facilities also drove a separation between design and manufacturing.[29]

The degree to which these two forms of vertical disintegration provided opportunities for de novo entrepreneurial start-ups can be measured by looking at the ratios of de novo to de alio firms among providers of standard modules and among design houses or foundries, and by measuring the increase in de novo firms among the largest U.S. semiconductor suppliers. To do this, each firm is defined as either de novo or de alio; as a design house, foundry, or integrated supplier of ICs; and as a supplier of standard modules or a broad-line supplier of either standard modules or nonstandard semiconductors. For the latter, more than 50 percent of sales must be from one type of standard module for a firm to be defined as a supplier.

None of the de novo or de alio firms that supplied semiconductors in the 1950s and 1960s can be classified as design houses, foundries, or providers of standard modules, and there were only 1 and 2 de novos, respectively, among the top 12 U.S. semiconductor suppliers in 1955 and 1965 (see Table 7.2). Beginning in the 1980s, however, the number of de novo firms among the largest U.S. semiconductor suppliers and among all standard module suppliers, design houses, and foundries increased considerably.[30] By 1995, 58 percent of the top 12 and 69 percent of the top 35 U.S. semiconductor suppliers were de novo firms (see Table 7.2); de novo firms also accounted for 59 of 64 providers of standard modules (92 percent) and 58 of 61 design houses (95 percent). As for foundries, their large economies of scale[31] enable only a small number to co-exist, and most of the leaders are in Taiwan. Furthermore, although it is not shown in Table 7.3, more recently founded de novo firms (after 1980) were more likely to be design houses or providers of standard modules than were those founded earlier (in or before 1980).

These results suggest that incumbents, whether de alio or de novo, that had been founded many years before the emergence of vertical disintegration had difficulties adjusting to it in the form of standard modules or a separation between design and manufacturing, and that these difficulties provided opportunities for new entrants. De alio firms had trouble adjusting to standard modules because these modules reduced the importance of their capabilities in system design, and they had trouble adjusting to a separation between design and manufacturing because this separation reduced the importance

TABLE 7.3

Differences between de novo and de alio semiconductor firms in 1995

	Standard modules versus broad-line or nonstandard semiconductors			Design houses, foundries, and integrated producers			
	Standard modules	Nonstandard semiconductors	Total firms	Design houses	Foundries	Integrated producers	Total firms
De novo	59	40	99	58	3	41	102
De alio	5	17	22	3	1	19	23
Total	64	57	121	61	4	60	125

SOURCE: Integrated Circuit Engineering (ICE; 1996).

NOTE: Foundries were not classified by "modules," and thus the total number of firms is different for columns 4 and 8.

of design-to-manufacturing integrative capabilities. Furthermore, the fact that older de novo firms also had difficulties adjusting to vertical disintegration suggests that vertical disintegration also reduced the importance of their integrative capabilities and thus provided opportunities for newly founded de novo firms. For example, increases in the number of transistors per chip (i.e., Moore's Law) not only drove improvements in standard modules but also continue to move the boundaries between them, which in turn necessitates changes in design rules and standard interfaces, thus providing further opportunities for de novo firms.

In summary, vertical disintegration and submarkets have enabled a large number of firms to co-exist in the U.S. semiconductor sector, and they have facilitated the success of de novo entrepreneurial start-ups. On the other hand, shakeouts have occurred in many of these vertically disintegrated layers because of economies of scale in R&D. For example, there are only a small number of foundries, and the largest ones (e.g., TMSC) are by far the most successful. Also, only a small number of firms dominate the market for ASSPs that target the PC (e.g., Intel and Nvidia) and the mobile phone (e.g., Qualcomm), and they are the most profitable.

We can also say that vertical disintegration between the semiconductor and electronic system sectors created new industries, by redefining the semiconductor sector as a separate entity. Design houses and foundries can be considered industries within the semiconductor sector because their products are based on a new concept, they are supplied by a new set of firms, and they represent billions of dollars in sales. Furthermore, some of the standard modules that resulted in vertical disintegration between the semiconductor and electronic system sectors can be considered new industries within the semiconductor sector as well because they are also based on new concepts (and thus defined as discontinuities in Chapter 6), supplied by new firms, and now represent revenues in the billions of dollars.

CONCLUSIONS

Understanding when and why the number of firms quickly declines (i.e., shakeout) is essential to understanding the challenges and opportunities of new industries. When economies of scale apply to a new technology in manufacturing, sales, or R&D, early entry is important for benefiting from them. When they apply to R&D, growing sales lead to increased R&D spending, and this spending can lead to new products and thus higher sales, and to smaller firms being acquired or exiting the industry. Generating this positive feedback between R&D spending, new products, and higher sales requires firms to release the right kinds of products at the right time. The analyses in Parts I and II help us better understand when new products that constitute a new industry might be economically feasible and thus should be released.

Large numbers of submarkets can reduce the impact of R&D economies of scale and thus prevent a shakeout. This is the case when different submarkets require different types of R&D. Since these submarkets often emerge over time as a new technology becomes economically feasible, they can enable firms to enter late and still survive. For this reason, understanding the timing of new submarkets is also important, and the issues covered in Parts I and II can help us understand when the new technology becomes economically feasible for them. For example, improvements in computer performance and cost, discussed in Chapter 4, expanded the number of applications for computers and thus the number of submarkets for new types of computers, software, and peripherals.

The emergence of vertical disintegration can enable firms to enter the markets for vertically disintegrated layers or even for the original product itself after a shakeout has occurred. For example, vertical disintegration in PCs provided opportunities in vertically disintegrated layers such as peripherals and software and in the PC itself as new entrants used the externally available software and peripheral modules to supply PCs. Looking more broadly at the IT and semiconductor sectors, this vertical disintegration enabled new firms to enter and continue entering, and it is still enabling new entries even now, more than 50 years after the initial emergence of these sectors. Although technological discontinuities also created opportunities, there would be far fewer opportunities if these sectors were still dominated by vertically integrated firms. For example, if electronic systems companies still produced their own electronic components, software, and sometimes peripherals, as many

Japanese firms do, far fewer opportunities would have been created in the U.S. semiconductor and IT sectors. Similar arguments can be made for broadcasting, financial services, and pharmaceuticals, where greater vertical disintegration in the United States than elsewhere facilitated the entry of new firms and kept these sectors globally competitive. Consider U.S. broadcasting (and movie distribution) companies. If they produced most of their own content, as many European companies did (e.g., Britain's BBC), far fewer opportunities would have been created in the broadcasting sector; in fact, opportunity creation is a one reason that some U.S. broadcasting firms are still global leaders.

These results suggest that recognizing how and when vertical disintegration and submarkets emerge is a critical issue for entrepreneurs. Just as understanding when new technological discontinuities might emerge is important for new and existing firms, so too is the emergence of vertical disintegration. There are some factors that drive both technological discontinuities and vertical disintegration. The timing of the discontinuities covered in Part II primarily depended on the performance and price of systems and their components; likewise, improvements in system performance facilitated the emergence of modular designs and thus vertical disintegration. For example, improvements in the processing power of mainframes and PCs facilitated the emergence of modular designs for the IBM System/360 and the IBM PC.[32]

However, there are significant differences between the timing of technological discontinuities and that of vertical disintegration. The timing of the latter also depends on factors such as when firms release open design rules; when these design rules are adopted by multiple firms and thus become standards; and when governments or other legal entities require firms to unbundle products, open their design rules, or reduce their ownership.[33] Entrepreneurs and incumbents that monitor these factors are more likely than those that do not to find new opportunities.

8

Different Industries, Different Challenges

This chapter explores the challenges that different industries present to firms, governments, and new industry formation in general. It does so by distinguishing between simple and complex systems and between industries that do or do not require a critical mass of users or complementary products for growth. The formation of most new industries depends on when a new technology becomes economically feasible and thus potentially provides a superior "value proposition" to an increasing number of users. Firms must recognize this potential and introduce relevant products in a timely fashion. Timing was addressed in Parts I and II.

However, industries represented by complex systems and/or that require a critical mass of users/complementary products to grow face additional challenges.[1] Overcoming these challenges often requires agreements on standards as well as new methods of value capture, new forms of industry organization, and new regulations. Some challenges may require government actions and so may delay industry formation. Depending on which countries can more effectively overcome them, these challenges may impact which countries first introduce and/or adopt a new technology and thus first experience industry formation.

A Typology of Industry Formation[2]

Figure 8.1 classifies industries in terms of complexity and whether a critical mass of either users or complementary products is needed for growth. The complexity of an industry can be measured in terms of the number of subsystems, processes, or lines of software code in the initial products or services

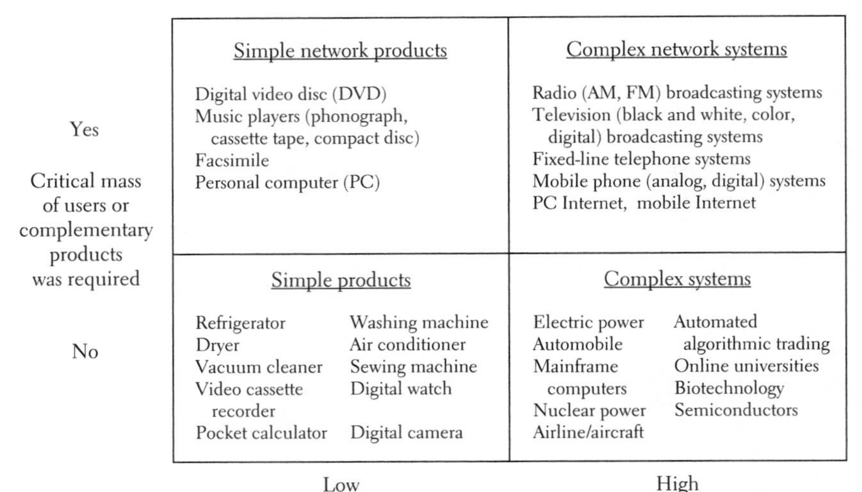

FIGURE 8.1 Typology of industry formation

it offers. In contrast to simple systems, complex systems usually require more technical, regulatory, and administrative decisions, and these may require the involvement of more organizations (including government agencies)—and often larger ones. Also, they may require more resources for their development (e.g., R&D).[3] Therefore, government support for R&D, government purchases and regulations, new forms of organization (including vertically disintegrated forms), and agreements among firms on standards and methods of value capture are more important for complex than for simple systems.

Whether a critical mass of either users or complementary products is needed for growth depends on whether a product by itself has value for users and/or whether there are immediate and large benefits to users from an expansion in either their own numbers or in the number of complementary products. Products such as telephones, facsimiles, video conferencing, short message services (SMSs), Internet mail, dating sites, and social networking, which exhibit direct network effects, required a critical mass of users to grow because value could only be obtained from them if there were multiple users and if those multiple users jointly used the products offered. On the other hand, products such as music and video players, some computers, and radio and television broadcasting initially required a critical mass of complementary products for growth because there were immediate and large benefits to users from expanding the number of complementary product lines. One reason for such benefits is that users demanded a wide variety of complementary products such as music, video, radio and television programs, and computer software. Furthermore, because the number of complementary products often increases as the number of users increases, the term *critical mass* as it applies to users is sometimes also used for products that display indirect network effects.[4]

However, the existence of network effects at some point in a product's life does not mean that the product initially required a critical mass of either users or complementary products to grow and thus for the industry to initially form. For example, consumers first used VCRs to record television programs, and the indirect network effects associated with prerecorded tapes did not emerge until after VCRs had begun diffusing.[5] And even if prerecorded tapes had not emerged, VCRs would probably have continued to diffuse, just as digital video recorders are now diffusing, because both products provide a recording function. Similar arguments can be made for Google, whose service benefits from indirect network effects between users and advertisers, in spite of the fact that users would benefit even if there were no advertisers. Other examples are mainframe computers and automobiles, which will be addressed momentarily.

SIMPLE PRODUCTS

The bottom-left quadrant of Figure 8.1 shows products of low complexity that did not require a critical mass of either users or complementary products for growth. The timing of the industries they represent depended on when these products offered a superior value proposition to some set of users and when firms recognized the potential for this superior value proposition and so introduced them. Because new products are often expensive, the first users are often those with high incomes; in fact, high income has been found to be the most important factor determining the initial country of diffusion.[6] This is one reason that most products representing new industries diffuse faster in the United States, Europe, and Japan than elsewhere. Furthermore, the United States had a much higher per capita income than Europe and Japan during the early post–World War II years, and so most new products first diffused there.[7]

Similar arguments can be made for capabilities. Many simple products (or simple network products) were first introduced by firms in the United States and Germany before World War II because these countries had greater capabilities in mass production and other technologies than did other countries. One example is electrical appliances industries, for which the United States and Germany had capabilities, including electrical technologies; also they were the first to introduce electric power.[8]

Examples of more recently introduced simple products include VCRs, handheld tape players (e.g., the Walkman), and digital cameras. These diffused first in wealthy countries, and Japan was initially their leading supplier because it introduced the right products at the right time, more so than other countries (as partly discussed in Chapter 5) and because it had previously developed the relevant capabilities. Japanese firms developed skills in magnetic tape through their early success in magnetic audio recording and playback equipment; these skills helped Sony and other firms dominate the market for handheld tape players. They have also been the leaders in digital cameras through their early development of relevant capabilities in analog cameras and ICs and in miniaturization in general—which they honed in a wide variety of products.[9]

SIMPLE "NETWORK" PRODUCTS

The difference between simple network products and simple products is in whether a critical mass of either users or complementary products is needed for growth. This depends on whether a product by itself has value to users and/or whether there are immediate and large benefits to users from an expansion in user numbers or in complementary products. For example, local area networks (LANs) and fax machines required a critical mass of users because multiple users were needed for value to be obtained from them.[10] A critical mass of users for both first emerged in firms that implemented them as a way to support intra-firm communication. The emergence of standards facilitated communication between firms and, in the case of fax machines, between consumers; however, rules against connecting fax machines to the public telephone network had to be eliminated before a critical mass of consumers could emerge.[11]

The other products shown in the upper left quadrant of Figure 8.1 required a critical mass of complementary products for growth. The value of new music, movie (e.g., DVDs), and computing formats often depends on the amount of software available for them, and thus there are large benefits to users from an increase in complementary products.[12] Each new type of music hardware—cylinders, shellac, vinyl, long-playing (LP) disks, cassette tapes, CDs (compact discs), and MP3 players—required music (i.e., software) to be made available for it.[13] Some firms supplied both hardware and software and thus created a critical mass of complementary products through a strategy of vertical integration. Other firms, in a strategy of vertical disintegration, focused on either one or the other, and their numbers grew as the variety of music exploded in the second half of the 20th century and as the music industry developed improved ways to agree on standards.[14]

Creating a critical mass of complementary products may have become more difficult in the last 30 years. Several new music formats, such as digital audio tape (DAT), digital compact cassettes (DCC), and minidisks failed in the 1980s and 1990s, and legal online music services have grown much more slowly than have other online services. As for DAT, DCC, and minidisks, their performance advantages over CDs were small and music companies did not strongly support them. DAT was not backward-compatible with cassette tapes, albeit DCC was, but the "switching costs" for CDs may have been too large for DCC to diffuse.[15] Similarly, online music sales lagged those in other online markets largely because the music companies did not support them. Not until Apple put together a new scope of activities and a revenue-sharing model did sales began to grow.

The PC industry may be an even better example of an industry that was temporarily immune to new discontinuities. Large switching costs and the ability of Microsoft and to a lesser extent Intel to control the "standard" interfaces between microprocessor, operating system, and application software in the so-called Wintel ("Windows-Intel") standard has probably slowed the introduction of new forms of computers whose operating systems and microprocessors are cheaper than Wintel's. Looking back to the 1970s, firms such as Apple and Tandy created a critical mass of complementary hardware and software for their computers, as IBM did in the early 1980s, by offering a more open platform that contained a hybrid 8- and 16-bit microprocessor. The 8-bit microprocessor made the IBM PC compatible with existing software, while the 16-bit microprocessor appealed to developers of new software and benefited from improvements in the cost and performance of microprocessors generally. Unfortunately, the openness of IBM's PC platform basically ceded control of the key internal interfaces to Microsoft and Intel, and this, along with the network effects associated with the Wintel standard, has probably delayed the emergence of a new PC-based industry.[16]

Only new discontinuities and an increasing compatibility between Wintel and Apple computers have reduced the importance of these network effects and thus the power of Microsoft and Intel. Many orders of magnitude in IC improvements have made this compatibility and discontinuities such as PDAs, tablet computers, and utility/cloud computing possible. These discontinuities have also reduced the importance of Wintel's network effects because PDAs and tablet computers appeal to new sets of applications and users.

The success of Palm and Apple highlights the different challenges associated with creating an industry consisting of simple products versus simple network products. With simple products such as VCRs, handheld tape players, and digital cameras, the challenge lies primarily in technical capabilities and in recognizing when a new product representing a new industry might become economically feasible. With simple network products such as PDAs, tablet computers, and MP3 players, however, capabilities for creating standards and alliances are also important. Part of Apple's (and to a lesser extent Palm's) success has to do with creating both standards for critical interfaces and alliances with media players such as music companies.

COMPLEX SYSTEMS

The systems shown in the bottom right quadrant of Figure 8.1 had high levels of complexity, but they did not require a critical mass of either users or complementary products to grow. Their initial products and services included a large number of subsystems, processes, or lines of software code, and this complexity increased the number of technical and policy decisions that had to be made (see Table 8.1). The quantity of decisions, the risks associated with pursuing multiple research avenues, and the sheer complexity of these systems meant that they required large amounts of R&D, that some governments provided large amounts of initial R&D funding, and that some governments were often the first customers for them. For example, government agencies

TABLE 8.1

Examples of critical choices in the early years of complex industries

Industry	Technical choice	Government policy
Electric power	Alternating current (AC) versus direct current (DC)	Right-of-way for power lines; safety regulations
Automobiles	Steam, electric, or ICE	Roads and safety regulations
Airlines/aircraft	Number of wings; location of engines and propellers	Military purchases and mail services
Computers	High-speed memory cache (mercury relay lines, cathode ray tubes, magnetic cores and drums) and processors	Purchases and R&D funding
Nuclear power	Pressurized-water reactors (PWRs) versus boiling-water reactors (BWRs)	Subsidies and R&D funding; safety regulations
Semiconductors	Material (germanium, silicon) and transistor (point-contact, junction)	Military purchases and R&D funding
Biotechnology	Method of screening and synthesizing drugs	Intellectual property (e.g., Bayh-Dole Act)
Automated financial algorithmic trading	Financial algorithm	Deregulation of financial markets
Online universities	Hardware or software platform	Subsidized loans for private universities

SOURCE: Hughes (1983); Kirsch (2000); Mowery and Rosenberg (1998); Flamm (1988); Cowan (1990); Tilton (1971); Kenney (1986); Pisano (2006); Patterson (2011); and Christensen, Johnson, and Horn (2008).

in the United States were the initial customers and/or the major funders of R&D for aircraft, airlines, computers, nuclear power, semiconductors, and biotechnology. This is a major reason for the early formation of these industries in the United States. The promulgation of appropriate regulations (or the elimination of regulations) impacted the formation of the electric power, airline, nuclear power, biotechnology, automated algorithmic financial trading, and online university industries, which is a major reason that these industries first formed in the United States.[17] Government purchases, R&D, and regulations played a much smaller role in the formation of industries represented by simple products and simple network products.

Consider, for example, the biotechnology industry, which began to emerge in the late 1970s and has continued to see strong growth in spite of its critics. New methods of synthesizing and sequencing drugs, which are based on advances in science, along with better computers, robotics, and now bioelectronic ICs, have enabled its growth. Biotechnology first emerged in the United States because of strong universities, a venture capital market, and a favorable regulatory system that protected intellectual property and enabled universities to benefit from federally funded research through, for example, the Bayh-Dole Act.[18]

Similar arguments can be made for automated algorithmic trading of stocks, bonds, currencies, and derivatives (e.g., credit fault swaps), which began to emerge in the 1970s and whose growth has been driven by improvements in both algorithms (i.e., advances in science) and computers.[19] Like biotechnology, algorithmic trading first emerged in the United States because of strong universities (i.e., finance departments), a venture capital market, and a favorable regulatory system, the last involving deregulation pursued by successive administrations. Online universities, too, benefited from a combination of improvements in computers and government-subsidized loans for private universities (i.e., policy), which supported the rapid growth of this industry more in the United States than in other countries.

On the other hand, none of the new industries shown in Table 8.1 required a critical mass of either users or complementary products for growth to occur. Each of the complex systems shown in the lower right quadrant of the figure was easily used in combination with other systems or was designed to stand alone. For example, automobiles used existing roads, fuel sources (e.g., cans of gasoline sold in general stores), and repair shops. Mainframe computers initially used existing punch cards in addition to other forms of secondary memory storage. Nuclear power stations used existing steam turbines and distributed their power over existing electrical distribution systems. The first semiconductors primarily replaced vacuum tubes in existing electrical systems. Some of these systems were initially stand-alones, including electric power for street lighting, mainframes, and airlines. In stand-alone systems, an integrated supplier provides all the hardware and complementary software that the user needs, and thus only supply-based economies of scale, as opposed to network effects that would require a critical mass of users or complementary products, were needed for these systems to grow.[20]

It is more difficult to say whether switching costs will emerge and cause these industries to confront the same kind of temporary barriers to new formation that were earlier described for the music and PC industries. Network effects did emerge for mainframe computers as a result of vertical disintegration. This enabled different firms to provide the computer, peripherals, and application software, but it did not prevent the emergence of the minicomputer, the PC, and the PDA industries. Network effects also emerged for automobiles in the form of gasoline and service stations,[21] and whether they will make it more difficult for electric or hydrogen vehicles to diffuse is unclear. Many automobile users may expect battery-recharging stations or hydrogen-refueling stations to be available, to the same extent that gasoline stations are available, before they will purchase automobiles that are powered solely by batteries or hydrogen fuel cells. For this reason, creating a critical mass of complementary products for these kinds of automobiles may be a major challenge.

COMPLEX "NETWORK" SYSTEMS

Complex network systems combine the challenges of complex systems with those of creating a critical mass of either users or complementary products (see Table 8.2). The difference between simple network products and complex network systems is that the latter involve more subsystems, processes, and/or lines of software code and thus require more resources (e.g., R&D), more decisions to be made, and often more organizations (including government agencies) to be involved with these decisions.

The difference between complex systems and complex network systems is that the latter require a critical mass of users or complementary products to grow. Complex systems whose value to users directly depends on user numbers exhibit direct network effects; those whose value to users directly depends on the number of complementary products exhibit indirect network effects. Creating a critical mass of users or complementary products and thus benefiting from direct or indirect network effects often requires more alliances, agreements on standards, and agreements on methods of value capture than those required by industries represented by complex systems.

For some of these industries, Table 8.2 summarizes whether they required a critical mass of users (direct network effects) or complementary products

TABLE 8.2

Examples of critical choices in the early years of complex network industries

Industry	Critical choices: By government	Critical choices: By firms	Critical mass required of:
Telephone	Government service versus private licenses Number of licenses Standards	Licensing of technology	Users
Radio and television	Government service versus private licenses Number of licenses Standards	Advertising versus subscription method of value capture	Complementary products (programming content and hardware)
Mobile phone	Number of licenses Choice and openness of standard	Choice and openness of standard Sale or rental of phones	Complementary products (phones and services for single standard)
PC Internet	Government and university R&D Degree of competition in telecommunications	Agreements on standards Initial focus on mail or file sharing	Users (mail) and complementary products (hardware, software, content)
Mobile Internet	Number of licenses Whether or not to push for open standards	Agreements on standards, method of value capture Whether to offer Internet mail	Complementary products (content, phones, services)

SOURCE: Brock (1981), Fischer (1987), Rohlfs (2001), Bussey (1990), Briggs and Burke (2002), Garrard (1998), Funk (2002, 2012a), and Kogut (2003).

(indirect network effects), and lists the types of firm and government decisions that supported industry formation. While telephones and Internet mail exhibited direct network effects, the other industries shown in Table 8.2 involved indirect network effects and thus their continued growth required a critical mass of complementary products such as hardware and software. Creating this critical mass required agreements on standards, the licensing of multiple firms, and new methods of value capture. Standards had to be set in the radio, television, mobile phone, and Internet industries,[22] and agreements on standards were necessary for a critical mass of hardware and software. Governments played an important role in promoting many of these agreements.

New forms of value capture also played an important role in these industries; this included advertising in radio and television, the subsidization of mobile phones by service providers, and broader and more complex methods of value capture in both the PC and the mobile Internet.[23] Finally, government licensing of multiple firms and in general promotion of competition impacted growth partly because finding the new methods of value capture required that multiple firms try a variety of different approaches.

Consider mobile phone systems and the Internet. Scandinavia experienced faster growth in mobile phone usage than did other countries, followed by the United States and Great Britain, because governments and government-sponsored entities more quickly created open standards, licensed service providers, and allowed domestic and foreign manufacturers to sell as opposed to rent phones. Scandinavian, U.S., and British service providers also set lower prices than those in other countries partly because the United States and Great Britain introduced more competition.. On the other hand, most Western European countries experienced faster growth in digital phones than did the United States because they chose a single standard and began licensing new entrants in the late 1980s, long before the United States licensed new entrants in 1995. This licensing was important because new entrants had more incentive to introduce digital services than did incumbent analog providers.[24]

Although the commercial formation of the Internet occurred in the early to mid-1990s, its seeds were planted in the 1960s through investments by the U.S. government. The government began funding the development and implementation of packet-switched networks that connected universities and other research institutions in the late 1960s, with a critical mass of Internet mail users first emerging among university researchers. This funding also enabled universities to set open standards for networks, browsers, and other technologies that expanded this critical mass. The government's early deregulation of the telecommunication industry and U.S. firms' early implementation of PCs and LANs contributed to the interlinking of government-funded, commercial, and corporate networks. The easy dissemination of browsers over these networks in the early 1990s and the lifting of restrictions on commercial activities in the mid-1990s triggered an explosion in Internet activity or, in this book's terms, the formation of the Internet industry.[25]

Japan and Europe (except Scandinavia and Great Britain) were much slower to fund Internet infrastructure, and when they did they focused on proprietary technologies that were supported by their "national champions." European and Japanese universities were also slower to implement information technologies partly because their governments funded less research and were slow to liberalize their telecommunication industries. This allowed other countries, particularly Korea, Singapore, and Hong Kong, to experience faster diffusion of Internet services than was seen in Japan and most of Europe.[26]

The mobile Internet experienced a completely different process of formation from that experienced by the PC Internet, in that firms, not universities, initially created it. Moreover, outside of Japan and Korea, these firms had trouble agreeing on standards for displaying content on phones and introducing other key services. Japanese service providers created a critical mass of users in 1999 and 2000 with phones that displayed content in a consistent manner (i.e., using standards), entertainment content that was supported by a micro-payment system, inexpensive Internet mail, and site access via URLs. They were able to obtain phones that displayed content consistently because, unlike Western service providers, they continued to dictate phone specifications to manufacturers. This also helped them introduce so-called micro-payment systems for these phones in which they charged subscribers for content and passed on about 90 percent of these revenues to content providers.[27]

Although Korean service providers also quickly created a critical mass of users, in a manner somewhat similar to that of the Japanese, Western service providers had trouble doing so because of a lack of agreement on standards, their unwillingness to make Internet mail a standard function on all phones, and their refusal to share a large percentage of revenues with content providers. Phone manufacturers failed to agree on standards in the WAP (Wireless Automation Protocol) Forum primarily because large ones such as Nokia believed that standards could lead to a situation like that found in the PC industry, where manufacturers' profits are much lower than Microsoft's and Intel's. For their part, service providers did not want to make inexpensive Internet mail a standard on all phones because they believed this would cannibalize their SMS revenues. These problems were finally solved by improvements in ICs and displays that made it possible for new entrants such as Apple and Google to introduce phones that could access PC content; this eliminated the necessity of creating a critical mass of mobile Internet content, users, and phones where the phones and content conformed to a single standard.

CONCLUSIONS

This chapter used a typology of industries to highlight the challenges of industry formation for different types of industries. The timing of formation for most new industries depends on when their products provide a superior value proposition to a growing number of users. However, those that involve complex systems or that require a critical mass of either users or complementary products face additional challenges. The difference between simple network products and simple products lies in the need to create a critical mass of either users or complementary products, the prerequisite for which is often agreements on standards. A critical mass of users had to be created in LANs and fax machines, and a critical mass of complementary products (hardware and software) had to be created for each type of music and video player and for computers such as PCs and PDAs. Standards played a critical role in the formation of these industries.

The development of complex systems typically requires more resources (e.g., R&D), more decisions to be made, and often more organizations (including government agencies) to be involved, than does the development of simpler systems. In many cases the increased costs of development (and the capabilities needed to effectively compete) required government R&D funding in, for example, electric power generation, mainframe computers, nuclear power, semiconductors, aircraft, and biotechnology. In some cases, high development costs and thus high prices meant that government-sponsored organizations such as defense departments or regulated monopolies were the earliest purchasers. Finally, the extent to which these complex systems impacted the use of land and other aspects of the environment led to the need for new laws or regulations that only government agencies could promulgate. Such laws and regulations were required for electric power generation, nuclear power, airlines, and biotechnology.

Complex network systems combine the challenges of complex systems with those of creating a critical mass of users or complementary products. Greater complexity increased the need for new laws and regulations, increased the costs of developing a system and thus the benefits from government purchases and support for R&D, and increased the need for agreements on appropriate methods of value capture for firms. The need to create a critical mass of users or complementary products increased the importance of making the right choices and obtaining agreements on them. In particular, an important task for governments was introducing competition while facilitating agreements on standards; for firms, it was making agreements on methods of value capture. Competition and agreements on standards were important for telephones, radio and television broadcasting, mobile phones, and both the PC and the mobile Internet.

IV
Thinking about the Future

Thinking about new industries and the future is inherently difficult. No matter how good a model is and now matter how well one does an analysis, there are new and unexpected customers, changes in consumer taste or government policies, rising incomes that lead to changes in consumer taste, unexpected advances in science, and the unexpected emergence of new concepts and architectures. This suggests that one must be careful about making forecasts. With this caveat in mind, Part IV uses the ideas from Parts I, II, and, to a lesser extent, III to think about new industries.

Because this book's characterization of technology change is very different from most characterizations, it uses a completely different approach to thinking about these new industries from that of most analyses.[1] Most analyses of potential new industries in, for example, books and documentaries focus on scientific and technical feasibility and, in particular, on fresh concepts in automobiles, trains, planes, ships, electricity generators, and the like, and their elegance and benefits. For consumer products, such analyses often assume that rising incomes will cause some technologies to diffuse even if their costs do not fall and their performance does not rise. Although there is some truth to this assumption, particularly for high-income consumers and the growing market for luxury goods such as private planes, yachts, and vacation homes, this book is more interested in technologies that can have a large impact on broad numbers of people, including those who are middle and lower income.

Consistent with the first three parts, Part IV focuses on a technology's potential for improvements in cost and performance and it uses the term *technological discontinuity* to distinguish new from old technologies. The chapters here analyze this potential using the four methods of improving performance and cost that were covered in Parts I and II. For a variety of new technologies, Part IV asks how their costs might fall and their performance might rise over the next five to ten years.

9

Electronics and Electronic Systems

Understanding when a technological discontinuity will become technically and economically feasible and thus begin diffusing is an important issue for firms, governments, and even students. Such an understanding can help firms better understand when they should fund R&D or introduce new products that are based on a discontinuity. It can help policy makers and analysts better understand when to begin developing policies and regulations, and it can help students better understand where to focus their careers.

Building from the overall notion of a technology paradigm, this chapter looks at when a variety of discontinuities became, or might become, technically and economically feasible by examining the improvements that these technologies experienced and/or currently are experiencing. Since key components play an important role in many of these technologies, this chapter largely draws on the interaction between components and systems, which was covered in Part II. For earlier discontinuities, it shows that many of the concepts and architectures that formed their basis were known long before they were implemented; instead, it was specific components that were the bottleneck to their commercialization. For discontinuities that are emerging or that may do so in the near future, this chapter shows how one can use performance and price data for systems and their key components to analyze when they might become technically and economically feasible. In this way one can distinguish between when the bottleneck is improvements in specific components and when it is a need for new concepts.

ELECTRONIC DISPLAYS[1]

Improvements in LCDs and other electronic displays have had a large impact on the performance and cost of electronic systems, probably larger than that of any other component except ICs. Both ICs and LCDs have been the primary drivers of improvement for radios, televisions, calculators, digital watches, audio systems, and digital cameras. The first LCDs, so-called passive-matrix displays, were used in pocket calculators and digital watches in the 1970s. The emergence of so-called active-matrix displays further expanded the number of applications, with improvements in semiconductor-manufacturing equipment facilitating this change.

In a passive-matrix display, an external circuit controls each pixel by successively supplying each row and column with an electric charge; the combination of row and column charges determines whether an individual pixel is on or off. In an active-matrix display, thin-film transistors (TFTs) determine the value of each pixel. Although much of the science for TFTs had been completed by the end of the 1960s, their implementation depended on the cost and performance of semiconductors and semiconductor-manufacturing equipment. When transistors were expensive, it was difficult to use TFTs, but as prices dropped the difference between the costs of active- and passive-matrix displays decreased. Furthermore, a small number of transistors, typically three, could together determine the choice of color for a specific pixel in combination with the appropriate materials for liquid crystals and polarizing light, so the cost of color displays fell as the cost of making TFTs on an active-matrix display fell.

Active-matrix LCDs were first used in high-end displays for applications such as fighter jet cockpits in the early 1980s. Solving yield and other manufacturing problems gradually enabled larger and cheaper displays to be constructed for applications such as laptop computers. As the demand for laptops increased, firms were able to increase the scale of LCD production equipment, and this led to dramatic reductions in cost, which still continue. (Costs were addressed in Chapter 3.)

Several discontinuities in displays are currently occurring. First, cold cathode fluorescent lights used as a backlight in LCDs are being replaced by LEDs; televisions with LEDs are sometimes referred to LED-backlit. Second, three-dimensional LCD televisions have been introduced and may diffuse over the next few years. Third, LCDs may be replaced by OLEDs in the next five years, assuming that their costs continue to fall and that their lifetimes can be lengthened. Fourth, looking farther into the future, holographic displays may become economically feasible in the next 20 years. Since the first and third discontinuities just mentioned were addressed in Chapter 2, this section briefly focuses on the second and fourth.

Three-dimensional television can be implemented using a number of methods, of which time-sequential 3D (with active LCD-based glasses) and auto-stereoscopic 3D (without glasses) are the most popular; both have been understood for decades. In time-sequential 3D, different images are shown to the left and right parts of a pair of glasses using frame rates fast enough to prevent blurring. By synchronizing the television and the LCDs in the glasses, an illusion of a 3D image is created. The bottlenecks for this discontinuity have been the slow frame rates of televisions and the cost and inconvenience of the glasses. The first bottleneck is being gradually overcome by improvements in frame rates, which were doubled between 2008 and 2010 because of lower-viscosity crystals and thinner crystal layers. These can be defined as components in an LCD whose improvements partly depended on finding materials that better exploit the relevant scientific phenomenon. The second bottleneck, the cost of LCD glasses is gradually being overcome by the falling cost of LCDs, while the third bottleneck, inconvenience still remains an issue; overcoming inconvenience may depend more on user attitudes than on technological change.[2]

Auto-stereoscopic 3D television relies on assigning different pixels in an LCD to the left and right eyes; an optical filter in the LCD ensures that each eye sees the correct pixels. While this method eliminates the need for glasses, it requires a very high pixel density to retain high resolution and to allow for viewer movements. This bottleneck is being gradually overcome by incremental improvements in pixel density (doubling every 1.5 years), which are being driven by the same factors that are driving reductions in IC minimum feature sizes. If the rate of doubling continues, it is expected that the required pixel densities will be reached by about 2016.

A final bottleneck for both time-sequential and auto-stereoscopic 3D is content.-Three-dimensional content is needed for users to benefit from 3D television, which is why it falls into the quadrant of "simple network products" that were discussed in Chapter 8. As with the other products in this quadrant, agreements on both standards and methods of value capture will be necessary. On the other hand, the falling cost of hard drives and ICs will reduce the cost of 3D content production.

Looking farther into the future, 3D holograms might become economically feasible as displays in the next 20 to 30 years. Although few people can describe or even define holography, many have seen the scene in *Star Wars* where a 3D hologram of Obi-Wan Kenobi speaks to Luke Skywalker. Many may also have noticed that 3D holograms are now used on credit cards. A critical difference between 3D holograms and the previously mentioned types of 3D television is that the former present an actual 3D image while the latter create an illusion of one. A 3D hologram is produced by lasers and holographic media, and a large hologram could be used in place of existing LCDs. Although a system that could produce a 42-inch image cost about $150,000 in 2011, it is expected that this cost will fall below $1000 over the next 20 years as the costs of computers (currently 60 percent of total cost), transmission media (10 percent), lasers (about 25 percent), and holographic media (5 percent) fall.

MOBILE PHONE SYSTEMS[3]

Improvements in the performance of mobile phone systems and their discontinuities have also been driven by improvements in components — in this case, electronic components such as vacuum tubes, analog and digital ICs, and displays. Although the origins of the wireless industry go back to the late 19th century, we simplify the discussion by starting with developments in the 1920s.

Improvements in vacuum tubes in the 1920s, which were driven by their use in music players and radio receivers, enabled the introduction of precellular phones, sometimes called private mobile radio, for military, police, fire, and taxi use. These systems used a single transmitter and low frequencies to cover a wide geographical area. The problem was that they could only handle a small number of users because of the limited frequency spectrum, which also caused the cost per user to be very high. The solution, which Bell Labs devised in the 1940s, was to divide a wide geographical area into cells and reuse the frequency spectrum in each one. The difficulty with implementing this solution at this time was that the components were both too expensive and too large. Low-cost ICs were needed for the fast and inexpensive switching equipment required to switch users to different base stations as they moved between cells. Low-cost ICs and low-cost analog components such as resistors and capacitors were also needed to reduce the cost and size of mobile receivers. By the late 1970s, digital switching was available, low-cost LED and LCD displays could be borrowed from pocket calculators, and receivers were small enough for an automobile. Scandinavia implemented the world's first systems in the late 1970s and early 1980s.

Further improvements in ICs drove the implementation of second-, third-, and now fourth-generation systems, which use the frequency spectrum for both voice and data more efficiently than earlier systems did. These systems include "air interface" standards that define how analog signals are converted to digital signals and how data is encoded and transmitted. The sophistication of the algorithms employed determines the capacity of the system, and more sophisticated algorithms require faster microprocessors and other digital chips in much the same way that digital audio and video players required advanced microprocessors (discussed in Chapter 5). For example, although time division multiple access (TDMA), which forms the basis for most second-generation systems, was conceived in the 1870s and applied to telegraph and later telephone systems in the 1960s, it could not be used in mobile phone systems until sophisticated ICs were available in the late 1980s. For third-generation systems based on wideband code division multiple access (CDMA), the time lag was a little shorter. Developed in the 1980s and first applied to satellite systems, narrow-band CDMA began to diffuse in the United States and a few other countries in the mid-1990s; wideband CDMA began to diffuse widely in the early 2000s as improvements in ICs made it feasible.

With continued IC improvements, engineers are now developing sophisticated algorithms that use the frequency spectrum even more efficiently and transmit data faster and more cheaply. These systems include wireless LAN and ultra wideband, which were briefly discussed in Chapter 2. Although new algorithmic concepts are needed and might partly depend on advances in science (e.g., new mathematics), designers consider the performance and cost of ICs when they choose an algorithm.

Another discontinuity that is quickly becoming economically feasible is cognitive or software radio.[4] While existing mobile phones and other wireless devices use electronic components that are dedicated to a specific frequency band and thus these devices are assigned a specific frequency, improvements in ICs and other electronic components are making it economically feasible for a single phone to use cognitive radio and thus access a wide range of frequencies and interface standards. This is important because allocating frequency bands is a time-consuming and expensive political process that causes wide swaths of frequency to be unused for many years, and because different countries use different standards.

Looking at the typology of industry formation from Chapter 8, cognitive radio can be defined as a complex network system that will require agreements on standards and changes in policies. The question that this chapter addresses is when cognitive radio might become economically feasible and thus when governments should begin thinking about standards and policies. Based on the number of transistors needed to implement cognitive radio on a single IC chip, and based on the falling cost per transistor, we can estimate when the cost of this chip will reach the cost of the chips that are currently used to access a dedicated frequency band (about $25) in a mobile phone. Such an analysis suggests that a single cognitive radio-based IC chip will be available before 2016. Similarly, the necessary improvements in filters and other electronic components are emerging, partly in the form of microelectronic mechanical systems (MEMS), which were mentioned in Chapter 6. Governments (and management and policy students) should begin placing more emphasis on this technology.

WIRELINE TELECOMMUNICATION SYSTEMS[5]

Improvements in telecommunication systems during the second half of the 20th century have involved a number of discontinuities, each of which is primarily the result of improvements in ICs. This includes the moves from electromechanical to electronic switches, from analog to digital systems, and from circuit- to packet-switched systems. In the move from electro-mechanical to electronic switches, crossbars and other mechanical assemblies were replaced by transistors and ICs as the cost of ICs dropped.

Digital systems initially used pulse code modulation, developed in the 1940s, and later used time division multiplexing (TDM), perfected in the 1980s. Although they could provide higher voice quality than analog systems, like digital music and video (Chapter 5), digital systems required much more data processing and thus depended on the availability of low-cost processors and other ICs. In addition to handling millions of 1s and 0s, more sophisticated ICs were needed to efficiently manage the different time slots in TDM.

Packet switching, a critical aspect of the Internet, was conceived in the late 1950s, improved in the 1960s, and demonstrated in the now famous ARPANET project in the 1970s. The economics of packet versus circuit switching in a digital system are quite simple. Paraphrasing MacKie-Mason and Varian[6] when lines were cheaper than switches, it was better to have more lines than switches and thus dedicate lines to each call or data transfer. In that way, each connection wasted transmission capacity (lines were held open whether or not data or voice signals were flowing) but minimized computer switching costs (one setup per connection). As computer switches became cheaper relative to lines, however, the network worked more efficiently if data streams were broken into small packets and sent out piecemeal, allowing the packets of many users to share a single line. Each packet had to be examined at each switch along the way to determine its type and destination, but relatively low-cost switches made this possible and enabled multiple packets to use the same, relatively expensive, lines.

The key term here is "relatively expensive." Although telephone lines were once relatively expensive, improvements in fiber-optic systems dramatically reduced the cost per call or packet of data, and thus the cost of transmitting large amounts of data, by many orders of magnitude. In a fiber-optic system, light is used to transmit information along a glass fiber, as opposed to electrical signals along a copper wire; the light is emitted by a semiconductor laser, sensed by a semiconductor photodiode, and amplified by ICs. In this way improvements in fiber-optic systems were driven primarily by improvements in these components.

For glass fibers, it was improvements in both the purity and the thinness of glass that reduced the cost of the fibers and enabled light to travel further before amplification was needed. Improvements in the purity of glass and in semiconductors enabled amplifiers to be placed on a glass substrate and thus reduced the need for converting light to electronic signals, which had previously reduced overall transmission speed. Third, the development of lasers that emit different wavelengths of light enabled systems to combine multiple lasers in a single glass fiber, using a technique called wavelength division multiplexing (WDM). Developed in the 1960s, WDM became economically feasible as improvements in lasers and control circuitry made it possible to send multiple wavelengths of light down a single strand of fiber, with each wavelength carrying different data.

Paraphrasing Gilder in his book *Telecosm*, the economics of telecommunication systems are being reversed: improvements in fiber-optic cables are outpacing computers. At some point in time the lines or the network will not be the bottleneck as characterized by MacKie-Mason and Varian in 1994. Gilder argues that computers will become the bottleneck and that there will be a scramble to develop an all-optical telecommunication and computer system. However, the difference between this bottleneck and others described in this chapter is that no clear trend tells us when such a system might become technically and economically feasible. The basic problem is that semiconductor lasers cannot be easily fabricated with silicon and thus cannot be easily combined with transistors and other optical components such as waveguides, couplers, resonators, and splitters on a single chip. Although these other optical components benefit to some extent from reductions in scale, just as ICs do, new concepts for making lasers on silicon are needed before such a system will become feasible either technically or economically, and these concepts might require advances in science.[7]

INTERNET CONTENT AND APPLICATIONS[8]

Improvements in Internet content and applications can also be analyzed in terms of improvements in components. In this case, the key components are telecommunication systems, as well as servers, modems, and PCs whose improvements are being driven by improved ICs. For example, improvements in the Internet made online universities economically feasible and continue to increase the economic viability of automated algorithmic trading, which was started by improvements in computers. Algorithmic trading has also depended on new, exotic financial products, each of which can be defined as a new concept that resulted from advances in science—specifically, mathematics.

On the consumer side, some improvements in modems made "streaming" of music possible and others made music and video downloads and streaming video possible. By the way, streaming can be defined as a new "concept" in which the time lag between its characterization and its commercialization was fairly short. Of course, all of this has changed with the diffusion of broadband services. Telephone and cable companies have used the improvements discussed in the previous section to increase bandwidth and upgrade systems from analog to digital, thus eliminating the need for modems. Combined with compression techniques such as MP3 and MP4, users with a 1-MB/second service can download a five-minute CD-quality song in seconds and a one-hour video in minutes. Combined with improvements in memory capacity and processing speeds in both portable and desktop computers, the Internet has become the main method of distribution for music and may soon become so for video. Increases in the memory capacity of desktop and portable devices are also making the memory requirements (about 128 MB for a one-hour video) rather meaningless.

These improvements, along with the improvements discussed earlier, are making three-dimensional (3D) content and applications possible. 3D movies, games, and other content, and applications such as flight simulation, oil field mapping, and digital special effects, will probably become technically and economically feasible online over the next few years. The speed at which such content and applications diffuse will depend to some extent on whether users will be willing to wear special glasses or will wait for auto-stereoscopic systems.

Another interesting application whose technical and economic feasibility is being driven by improvements in telecommunication systems is the mashup, a web application that combines data from more than one source into a single integrated tool. To be effective, a mashup should involve easy and fast integration and open APIs (application programming interfaces) and data sources, and it should enable results that may not have been the original goal of the data owners. For example, eBay and Amazon.com have enabled many of their users to integrate their applications in order to reduce the costs of managing databases, writing new pages, and entering data. According to *Wikinomics*, in 2007 40 percent of the goods listed on eBay were being automatically uploaded from the inventory systems of third-party stores that use eBay as an alternative sales channel. Similarly, Amazon enables 140,000 software developers to access its product database and payment services to create their own new offerings.

An example of a more sophisticated mashup is Google Earth. As many readers know, Google has opened up many of its APIs, including Google Earth, enabling other sites to build their services on top of them and thus avoid the costs of creating their own databases. For example, the site Housingmaps combines output from Craigslist and Google Earth to enable users to look for housing on a map of the relevant area. Since most home searchers are interested in location, pinpointing houses and apartments on a map, particularly a sophisticated one like Google Earth, provides many potential benefits, as improvements in Internet connections, PCs, and telecommunication systems that allow fast site navigation have occurred in the last few years as this book was being written.

UTILITY/CLOUD COMPUTING[9]

The Internet improvements described in the last two sections are also driving changes in system architectures for both businesses and consumers which have already undergone many changes in the last 50 years. Chapter 4 mentioned some of these changes and how improvements in components have altered the trade-offs that users make when they consider information systems. In the 1960s and 1970s, firms made trade-offs between owning a computer system and accessing remote services through inexpensive terminals. The high price of computers initially caused small firms to choose the latter. Improvements in components, particularly in ICs, changed these trade-offs and enabled users to purchase PCs, which eliminated the need for remote services in the 1980s. Now improvements in the Internet and the increasing complexity of software are making remote services, in the form of utility/cloud computing, economical again, as Carr describes in *The Big Switch: Rewiring the World, from Edison to Google*.[10]

In utility/cloud computing, computing services are provided over the Internet in a manner similar to how services such as electricity, water, and gas are provided by utilities. These services include software, data storage, platform, and infrastructure. Improvements in the Internet are making these services economical because they reduce costs (of transmission and the like) and increase benefits (e.g., access from anywhere).[11]

Consider expensive software, such as enterprise resource planning (ERP), customer relationship management (CRM), and sales force automation (SFA), that is widely used by large organizations. Because the implementation of this software typically costs many millions of dollars,[12] small firms and certainly individual users have never been able to afford it and instead have been the first users of software as a service (SaaS) and other utility/cloud computing offerings that have very low implementation costs. Small firms can now manage inventory, customers, and sales information in a manner that may be inferior to what large firms can do with customized applications, but they can do it for a fraction of the cost.

As providers of SaaS increase the variety of their services and as improvements in the Internet continue, the size of the organization for which SaaS is economical will also increase. A greater variety of services will reduce the gap in customizability between SaaS and packaged software, and Internet improvements will enable better security and reliability—two of the most cited barriers to implementing SaaS and other aspects of utility/cloud computing. Furthermore, as the providers of utility/cloud computing continue to grow, they will be able to more effectively solve the problems of security and reliability than their clients will.[13] Solving other widely cited problems such as lock-in and control[14] may depend more on strategic and organizational factors than on improvements in technology.

MOBILE INTERNET CONTENT AND APPLICATIONS[15]

Improvements in mobile phone networks, which were discussed previously, and in digital ICs and displays had made a simple form of the mobile Internet technically and economically feasible by the year 2000, and they continue to make more sophisticated content possible. Initially, however, demands for lower power consumption, lower prices, and smaller phones severely restricted a search for potential applications. It was not possible to download or even process music, video, dynamic content, color pictures, or even regular Internet mail on phones in 1999 and 2000. Instead, it was simple musical chords that could be used as ring tones; simple pictures that could be used as wallpaper or screen savers; and text-based sites, short mail messages, and later simple games drove the mobile Internet's initial growth. An unexpected requirement was a micro-payment system that would allow content providers to charge users for content.

As discussed in Part III, Japanese service providers were the first to successfully offer these services. While Western firms initially focused on sophisticated content, the Japanese were first with micro-payment services, entertainment content, and, most important, standards. Moreover, they were the first to convince manufacturers to supply them with phones that corresponded to these standards. Downloading ringtones, wallpaper, and text requires compatibility among phones, content providers, and services in order to create a critical mass of users. Furthermore, within a single phone there must be compatibility between the way text, pictures, and ring tones are formatted, users are authorized, and payments are made. Although compatibility is not difficult when processing speeds and memory capacities are high, and thus tasks can be carried out in a decentralized modular manner, this was not the case in 2000. Ensuring this compatibility required decisions to be centralized by a single firm that carried out so-called integral design.

Nevertheless, improvements in processing speed, memory capacity, and display, and in mobile phone networks, continued as better ICs become available. In the early 2000s, these improvements enabled higher-quality ring tones and later CD-quality music, higher-quality wallpaper (including the use of camera phones to create it), higher-quality games and video, and PC-Internet content compatible browsers. PC-Internet content compatible browsers made it possible for Internet content to be downloaded and thus eliminated the need to develop a critical mass of mobile Internet content along with a critical mass of phones and services to support it, which only the Japanese and Koreans had managed to do.[16]

Further improvements in processing speed, memory capacity, display, and networks will continue to expand the capabilities and importance of the mobile phone. They will also continue to improve the performance and cost of applications such as physical payments, videophone calls, location-based services, mobile televisions, and infrared and Bluetooth connections with other devices, and they will enable most PC-based applications such as 3D content, mashups, SaaS, and other aspects of utility/cloud computing. Understanding when specific applications might become technically and economically feasible, which is the purpose of this book, can help governments, firms, and even students identify the timing for these applications and then analyze strategies and policies for them.

HUMAN-COMPUTER INTERFACES[17]

Human-computer interfaces are an important part of almost every electronic system. Until recently, most were graphical user interfaces that consisted of a keyboard, a mouse, and the appropriate display, but now touch screens are technically and economically feasible partly because of lower costs of depositing LCD-related materials and forming patterns in them, and better control over the thickness and other layer characteristics.

A another possibility for human-computers interfaces is automated speech recognition, made possible by improvements in ICs and the signal-to-noise ratio of microphones However, the error rates are still too high for most applications, except for those involving reading a speech or using an air travel kiosk, but they are falling fairly quickly as ICs are improved. In any case, many of us would only use a limited number of words and speak in a clear, even voice in attempting to direct a computer.[18] Whether we will be comfortable speaking to our computers in a public or private area is a different question.

A third possibility for interfaces is gesturing. First made economically feasible by microelectronic mechanical systems (MEMs) in a Wii portable game, gesturing may be possible in other computer applications if gestures can be interpreted with cameras. This will require a camera with very high spatial (number of pixels) and temporal (high frame rate) resolution, light sensitivity (high pixel sensitivity), depth sensing (low depth error), and affordability. One reason that light sensitivity is important is that as pixel size is decreased, the amount of light available for each pixel declines. Improvements of about 100 times have been made in this area since 1986[19] and will continue to be made in both spatial resolution and sensitivity and in the other dimensions of performance, largely because of the same factors driving Moore's Law.

The ultimate in human-computer interfaces is a neural interface, which senses the magnetic fields generated by our brains as we take in and process information. In this way we will probably be able to control computers much more quickly and effectively than with other interfaces. Although many of these developments have been rightly directed at users with physical handicaps, a neural interface may be of interest to a broad range of users if the price is sufficiently low and the spatial and temporal resolutions of the magnetic fields are sufficiently high. While these resolutions were improved by more than five magnitudes between 1975 and 2009, two more orders of magnitude are probably needed, as well as dramatic reductions in the current price of $30,000 per system. Both may require many new concepts, which may depend on advances in science.

CONCLUSIONS

This chapter looked at when a variety of discontinuities became, or might yet become, technically and economically feasible by looking at improvements in them. Since key components play an important role in many of these technologies, this chapter largely built from the interaction between components and systems that was covered in Part II.

For past discontinuities, the chapter showed how many of the concepts and architectures that formed their basis were known long before the discontinuities were implemented; however, specific components were obstacles to their technical and economic feasibility. Partial exceptions include audio and video streaming, podcasting, blogs, social networking, and mashups, which were developed in response to improvements in ICs and the rapid growth of the Internet. For these it may be that the emergence of new applications occurred much faster than expected and thus it took some time for scientific research to catch up with the demands of the marketplace. As discussed in Chapter 6, something similar occurred with semiconductors in the 1950s and later with new forms of memory such as EPROMs, EEPROMs, and flash memory.

For future discontinuities, this chapter showed that performance and price data for systems and their key components can be used to analyze when they might become technically and economically feasible. We can distinguish between when improvements in specific components are the bottlenecks to economic feasibility and when new concepts are needed that may depend on advances in science. The rate at which component improvements occur can tell us roughly the rate at which various forms of 3D displays, cognitive radio, mashups, cloud computing, new mobile Internet-based services, and gesture-based human-computer interfaces will become feasible, both technically and economically. For other discontinuities such as optical computing, neural interfaces, and perhaps 3D holograms, new concepts may be required, which may depend on advances in science.

10

Clean Energy

It is widely recognized that the world needs to decrease its use of fossil fuels in order to reduce carbon dioxide emissions. Although reducing carbon dioxide and other greenhouse gases requires changes in lifestyle, cleaner energy technologies such as wind turbines, solar cells, and electric vehicles are also an important part of any solution to global warming. Building from the overall notion of a technology paradigm, this chapter shows how this book's interpretation of technological change and industry formation can assist engineers, managers, policy makers and analysts, professors, and even students in better analyzing the potential for clean energy and in devising better policies.

Indeed, the technology paradigm can be of more help than theories of technological change that currently form the basis for government policies in many countries. These policies emphasize demand-based over supply-based subsidies. One estimate is that 95 percent of the U.S. budget for clean energy is for demand, leaving only 5 percent for supply-based subsidies such as funding of R&D.[1] A major reason for this is that many believe that costs will fall as more wind turbines, solar cells, and electric vehicles are produced and that this "learning" primarily occurs within the final product's factory setting as automated equipment is introduced and organized into flow lines.

This chapter shows that, for wind turbines, the cost of electricity generated has primarily fallen as their scale has been increased, that further increases in scale require better "components" such as materials with higher strength-to-weight ratios in turbine blades, and that likely limits to increases in scale suggest a need for new designs. For solar cells, although their production does matter, it is more about increasing the scale of both substrates and production equipment than about simply automating equipment. More important, it is about improving efficiencies and reducing the thickness of active materials. For electric vehicles, again it is not production that primarily matters; it is the technology paradigm for their key component (the energy storage device) and whether and how its energy storage density (e.g., batteries or something better in the future) and costs (through, for example, larger-scale production equipment) can be improved.

This book's interpretation of technology change has important implications for government policy in that supply-based policies would probably stimulate the necessary improvements in cost and performance more effectively than would current demand-based policies. Since the latter reward firms for production, they have encouraged continuation of existing technologies—horizontal-axis wind turbines, crystalline silicon solar cells, and hybrid vehicles with lithium-ion batteries—in spite of data suggesting that other technologies probably have a greater potential for improvements and some already have lower costs..

WIND TURBINES[2]

Wind turbines translate spinning blades into electricity using a generator, just as conventional electrical generating stations do. The biggest difference is that the wind turns the blades and thus the generator in a wind turbine while steam (say from burning fossil fuels) turns a generator in conventional electricity generation. For this reason, wind turbines must be placed in locations with high wind speeds, which are often far from homes and factories, and new transmission lines are often required to connect so-called wind farms to users. To simplify the analysis, this section focuses on the cost and output of the wind turbine itself partly because the cost of a wind turbine typically contributes about 75 percent to the capital cost of a wind farm.

Although there are a number of wind turbine designs, the horizontal axis has been used in almost all new installations since the early 1980s. This is basically a tower with a three-blade rotor (which looks like a propeller) mounted at the top, where the diameter of the rotor defines the "swept area of the blades." Theoretical and empirical analyses suggest that the largest source of reductions in the cost of electricity generated from wind turbines is geometrical scaling, which works in a number of ways.

First, the theoretical output of a wind turbine is a function of diameter squared (the area swept by the turbine blades) and wind speed cubed (see Table 10.1). Thus, increases in rotor diameter can lead to lower cost per output if the cost of the rotor blades does not rise as fast as the output of the turbine. Second, empirical analyses of rotor diameter and power output for actual installations have found that output is a function of diameter to the 2.254th power (see the second equation in Table 10.1), or slightly greater than the theoretical prediction of diameter squared. This is probably because (actual data on wind speed was not available for these installations) larger wind turbines can handle higher wind speeds than smaller ones can and so benefit from the fact that output is a function of wind speed cubed. One reason that larger turbines can handle higher wind speeds is that, technically, they have a higher maximum-rated wind speed than smaller ones, which shut down at a lower speed (see the third equation in Table 10.1). A second reason is that wind velocity is often lower near the ground because of uneven terrain or buildings; the higher the tower, the higher the wind speed, and thus the higher the power output (see fourth equation in Table 10.1).[3]

Empirical analyses of cost also suggest that there are benefits from geometrical scaling. Costs do not rise as fast as output as both swept diameter and tower height are increased, albeit the limits of scaling are still uncertain.[4] For the tower, analyses have found that cost is a linear function of height, with small increases particularly beneficial (see the fifth equation in Table 10.1). For example, increasing height from 10 to 20 meters causes output to rise by 94 percent and costs to rise by 9 percent. This suggests that higher towers, which are often needed for larger rotors, can lead to lower costs of electricity.

Turning to turbine blades, costs are usually analyzed in terms of area swept—in other words, cost per rotor area. Empirical analyses have found that for diameters less than 50 meters, these costs have actually fallen as rotor diameter has increased (see the sixth equation in Table 10.1). Only for

TABLE 10.1

Relevant equations for wind turbines

	Phenomenon	Equation
1	Theoretical output	Power output = $3.229 * (\text{diameter})^2 * (\text{wind velocity})^3$
2	Best-fit regression equation of installations	Power output = $0.1034 * (\text{diameter})^{2.254}$
3	Best-fit regression equation of installations	Maximum-rated wind speed = $9.403 * (\text{diameter})^{0.081}$
4	Theoretical output and best-fit regression equation of installations	Power output = output at ground $* (\text{tower height})^{0.96}$
5	Best-fit regression equation of installations	Tower cost = $0.85 *$ cost of steel $* (\text{diameter})^2 * (\text{tower height}) - 1414$
6	Best-fit regression equation of installations for diameter less than 50 meters	Cost/output = $434 * (\text{diameter})^{-0.328}$
7	Best-fit regression equation of installations for diameter greater than 50 meters	Cost/output = $935.2 * (\text{diameter})^{0.103}$

diameters greater than 50 meters have costs risen, and they have actually risen slightly faster than the output of actual installations. This means that if one ignores the benefits from higher wind speeds, the cost per output has actually risen as diameters have increased beyond 50 meters (see the seventh equation in Table 10.1). Presumably, if one were to include wind speed in the analysis of output (phenomenon 2 in Table 10.1), increases in rotor size beyond 50 meters would result in a decline (albeit slow) in cost per output. We come to this conclusion because most wind turbines being installed today have rotor diameters near or beyond 100 meters.

In any case, these analyses provide further evidence that the current emphasis on cumulative production as the main driver of cost reductions is misleading, and they provide better guidance about appropriate policies. If the cumulative production of wind turbines were driving cost reductions, manufacturers would produce more small turbines rather than a few large ones to meet some fixed amount of capacity. Nevertheless, we may be approaching the limits to increased scale of horizontal-axis wind turbines, and these analyses may help us in overcoming them. The costs of the turbine blades are quickly rising as rotor diameter increases, largely because larger rotors require more expensive materials. Aluminum, glass fiber–reinforced composites, and wood/epoxy can be used for small diameters, but carbon fiber is needed for diameters of greater than 50 meters.

Carbon fiber–based blades have higher strength-to-weight ratios, which are needed to reduce the stresses on both the blades and the adjoining components, such as the gear box, the turbine, and the tower. The problem is that these blades have much higher costs than older blade types, which can be manufactured with methods borrowed from pleasure boats such as "hand layup" of fiberglass reinforced with polyester resin in an open mold. Carbon-based blades require better manufacturing methods, such as vacuum bagging and resin infusion, that have been borrowed from the aerospace industry. This suggests that a combination of higher strength-to-weight ratios and lower-cost materials are needed.

While subsidizing the installation of wind turbines will lead to more turbines and perhaps ones with larger rotor diameters, it may not lead to lower costs per output without new materials. And these materials will probably not emerge unless R&D spending on them is significantly increased, partly because some of them may depend on advances in science. Only a small portion of demand-based subsidies end up funding R&D, whereas supply-based subsidies go directly to R&D.

Research on ultra-strong adhesives so that large blades can be assembled from small pieces onsite would also be potentially valuable. This is because transporting blades longer than 50 meters is a large problem and solving it may represent as much as 20 percent of the total cost of installation.[5] A 145-meter-diameter wind turbine is being constructed in Norway[6] that will be larger than the London Eye (122 meters) and almost as large as the Singapore Flyer (150 meters) Ferris wheels, so these installation costs may continue to rise.[7] The good news is that ultra-strong adhesives are currently being used to assemble airframes for aircraft and to some extent can be borrowed from the aircraft industry for the assembly of turbine blades.

If we are approaching the limits of scaling for the horizontal-axis wind turbine, radical new designs may be needed. This suggests more funding of R&D and fewer subsidies for installation. For example, a British firm has proposed a V-shaped design in which two additional blades are attached perpendicularly to the ends of two main blades to form a V. This design reportedly weighs half of the conventional design, and it can be used in offshore locations, where wind velocities are higher and often more consistent, even at the ocean surface. Also, it does not require tethering to the ocean bottom and thus can be installed in deep water. The manufacturer claims that it will introduce a 275-meter diameter version by 2014.[8]

Others have proposed even more radical designs, such as attaching wind turbines to balloons, kites, or other buoyant objects. Such designs would enable access to higher-speed winds, and their use of light and flexible materials might enable them to handle higher wind speeds than can terrestrial-based turbines. Some estimate that the cost of electricity from these turbines could be as low as $0.01 per kilowatt-hour, or much lower than the current cost of electricity.[9]

With both tethered and V-shaped wind turbines, supply-side policies such as greater spending on R&D are probably more appropriate than demand-side policies, which have merely encouraged existing designs, albeit those that have larger scale. Demand-side policies have not encouraged development of new designs, as others have noted,[10] and new designs are probably needed before the cost of electricity from wind falls below that of conventional sources.

In summary, both the need for new designs and the need for new materials to benefit from increasing the scale of existing wind turbines suggest a need for supply-side policies. A technology paradigm can help us identify these policies much more so than the conventional wisdom that costs fall as cumulative production increases.

SOLAR CELLS[11]

Most solar cells translate incoming solar radiation into electricity via the photovoltaic effect. This effect was first recognized in 1839 by the French physicist

Alexandre-Edmond Becquerel; the first solar cell was constructed in 1883 by Charles Fritts; the modern junction semiconductor solar cell was first patented in 1946 by Russell Ohl; and the first silicon solar cell was constructed in 1954 by Calvin Fuller, Daryl Chapin, and Gerald Pearson.

According to the photovoltaic effect, incoming solar radiation creates "electron-hole" pairs in an appropriate material which create electricity when they reach the opposite terminals of the device. Only photons whose energy (which is higher for lower wavelengths of light) exceeds the band gap of the material create electron-hole pairs, and the energy that results from these pairs is equal to the band gap. Thus, there has been a search for materials that exploit the photovoltaic effect, that have the appropriate band gap, that have little recombination of electrons and holes before the pairs reach the terminals, and that are inexpensive to acquire and process.

The first three factors affect efficiency; the fourth affects cost per area. It is the combination of efficiency and cost per area that determines the cost per peak watt, which is the measure ordinarily used to evaluate solar cells. The final cost of electricity from solar cells roughly depends on this cost along with the degree of incoming solar radiation and the cost of installing the solar cells and their related electronics. The degree of incoming solar radiation is highest in countries near the equator and in those with low cloud cover.[12] The cost of installing solar cells depends on a number of factors such as increasing the scale of a solar "farm" and the ability to use solar cells as part of a building's structure. To simplify things, this section focuses on efficiency and cost per area and thus on cost per peak watt, which has fallen over time.[13]

Efficiency

The efficiency of solar cells is measured in terms of the percentage of incoming solar radiation that is translated into electricity. The factors that impact efficiency can be divided into those relating to maximum theoretical efficiency, the best efficiencies produced in a laboratory, and the best efficiencies produced in mass-production facilities. The first factor that impacts maximum theoretical efficiency is the band gap of the material. Since only photons with energies greater than this band gap can create electron-hole pairs, materials with lower band gaps increase the percentage of incoming radiation transformed into electrons and holes, but they reduce the energy from each absorbed photon. However, materials with higher band gaps increase the energy from each absorbed photon but reduce the percentage of incoming radiation that can be transformed into electrons and holes. Thus, there is a trade-off between low and high band gaps, and, given the distribution of the solar spectrum,[14] there is an optimum.[15] The second factor that impacts theoretical efficiency is the extent of recombination of electrons and holes before they reach the terminals. Because of an easier flow of electrons and holes, single-crystalline silicon cells have less recombination of electrons than, say, polycrystalline and amorphous silicon cells.

One way to overcome the limitations of individual materials is to use multiple "junctions," each with a band gap that is appropriate for a different part of the solar spectrum. Multiple-junction solar cells can theoretically be much more efficient than those with single junctions, but they also have higher costs because multiple layers must be deposited, patterned, and etched, and they often use exotic materials such as gallium arsenide. Incoming sunlight is often focused on multi-junction cells using a lens or a so-called concentrator to reduce the amount of photovoltaic material needed to produce a given amount of electricity. Concentrating mirrors currently require mechanical and electronic controls, gears, and other potentially unreliable components, and they can only be used when there is a great deal of direct sunlight, such as in cloudless skies.[16]

Achieving theoretical efficiencies in a laboratory or in a production facility is a major challenge. In a laboratory, efficiencies have depended on a wide variety of changes that include finding the right materials for the individual layers, achieving the necessary levels of purity in materials, and implementing the best processes, including the best temperatures and pressures. These advances have led to smooth but diminishing improvements,[17] which is consistent with other technologies covered in this book. In a production facility, better efficiencies require better process consistency, and this requires better processes and equipment, both of which partly depend on a better scientific understanding of the interaction between materials, processes, and efficiencies.

Table 10.2 summarizes the theoretical efficiencies and the best laboratory and production efficiencies for selected solar cell types. Since the three-junction cell with concentrators uses very expensive materials and is only employed in applications such as satellites, where weight and thus efficiency are much more important than costs, this technology will be ignored. As for the other technologies, the first interesting thing about them is that the efficiencies achieved in production facilities are much smaller than the theoretical efficiencies. This suggests that there is a large potential for improving the efficiencies of working solar cells and thus the cost per peak watt. Other than with crystalline silicon—the most widely used and most mature technology—the best production efficiencies are still less than half their theoretical counterparts. Furthermore, it is important to recognize that efficiencies have a much

TABLE 10.2

Best solar cell efficiencies and theoretical limits

Technology	From production facilities (%)	From laboratories (%)	Theoretical limits (%)
Three-junction cell with concentrators	25	42	63
Two-junction amorphous and microcrystalline silicon cell	NA	22	32
Crystalline silicon	18	25	29
Microcrystalline silicon	14	20	29
CIGS	11	20	29
CdTe	11	17	29
Amorphous silicon	8	13	20
Organic cells	2	8	31
Dye-sensitized cells	NA	12	31

SOURCE: U.S. Department of Energy (2010) and Wang Qing and Palani Balaya (personal communication).

NOTE: CIGS, cadmium indium gallium selenide; CdTe, cadmium telluride.

larger impact on the final cost of electricity than just their direct impact. In other words, doubling the efficiency of a solar cell can reduce the cost of electricity by more than one-half because the efficiency of a module, which combines multiple solar cells, is only as high as the least efficient cell in it. Thus, any variation in solar cell efficiency, which can come from poor process controls in a factory, has a large impact on the efficiency of a module and thus on the cost of electricity from solar cells.

To what extent might production facilities be able to produce solar cells with efficiencies close to their theoretical limits? This is a difficult question. Since crystalline silicon has received far more research and investment than have other solar cell materials, it has higher ratios of "best production" to theoretical efficiencies. This suggests that as long as more research dollars flow toward other solar cell materials, it will probably be possible to increase their ratio of best laboratory to best production efficiencies to the same levels seen with crystalline silicon. Furthermore, improvements in production processes, which are partly a function of cumulative volume, can probably enable efficiencies for mass-produced cells to approach the best laboratory efficiencies. This suggests that efficiencies of at least 20 percent for silicon, CIGS (cadmium indium gallium selenide), CdTe (cadmium telluride), and dye-sensitized cells, which would be almost a doubling for CdTe and CIGS and even more for dye-sensitized and organic cells, are possible. As argued next, supply-side policies would likely facilitate these improvements more so than demand-side subsidies.

Costs per Area

The cost per area of solar cells depends on the costs of materials and the equipment needed to process them. The cheapest materials are those for organic or dye-sensitized cells. Like organic light-emitting diodes (OLEDs), described in Chapter 2, organic cells rely on inexpensive polymers for which the relevant patterns can be roll-printed using conventional printing techniques, as opposed to the high-temperature processes used for semiconductor materials. Similarly, photosensitive dyes can be made from inexpensive materials such as titanium dioxide, which is also used as an additive for paint. As mentioned earlier, further reductions in cost per peak watt for these solar cells can be achieved by increasing their efficiencies, for which there is a large potential.

With each technology, one way to further reduce the cost of materials is to use thinner layers of them, which is a form of geometrical scaling. For example, producers of crystalline silicon solar cells cut thinner wafers from a cylindrical silicon ingot with laser-cutting machines, which also reduces the amount of wasted silicon. However, much greater reductions in material thickness have been and still can be achieved using so-called thin-film solar cells. These include organic, dye-sensitized, amorphous silicon, CIGS, and CdTe cells. While the thinness of a single-crystalline or polycrystalline cell has been limited by the ability to slice a round wafer from a cylindrical silicon ingot, the thinness of the other types of cells partly depends on an ability to deposit thin films onto inexpensive substrates using chemical, physical vapor, and sputtering techniques that were perfected for semiconductors and LCDs. The thinness of these layers is a major reason that thin-film cells have much lower cost per area and even cost per peak watt than do single-crystalline and polycrystalline cells. As shown in Figure 10.1, the costs per peak watt of thin-film amorphous silicon, and even more so for CIGS and CdTe, are now much lower than the costs of polycrystalline silicon (referred to as multicrystalline in Figure 10.1).[18]

In spite of these lower costs, however, more than 80 percent of solar cells produced in 2010 were based on crystalline silicon and most thin-film cells were based on thin-film amorphous silicon.[19] Single-crystalline and polycrystalline solar cells are emphasized more than thin-film cells, and even thin-film amorphous silicon cells are emphasized more than other thin-film cells because it is easier to build a factory to produce them and to obtain demand-based subsidies. Turnkey factories and equipment for single-crystalline, polycrystalline, and even amorphous silicon cells are much more available and have been for much longer than those for cells using CIGS and CdTe. For the

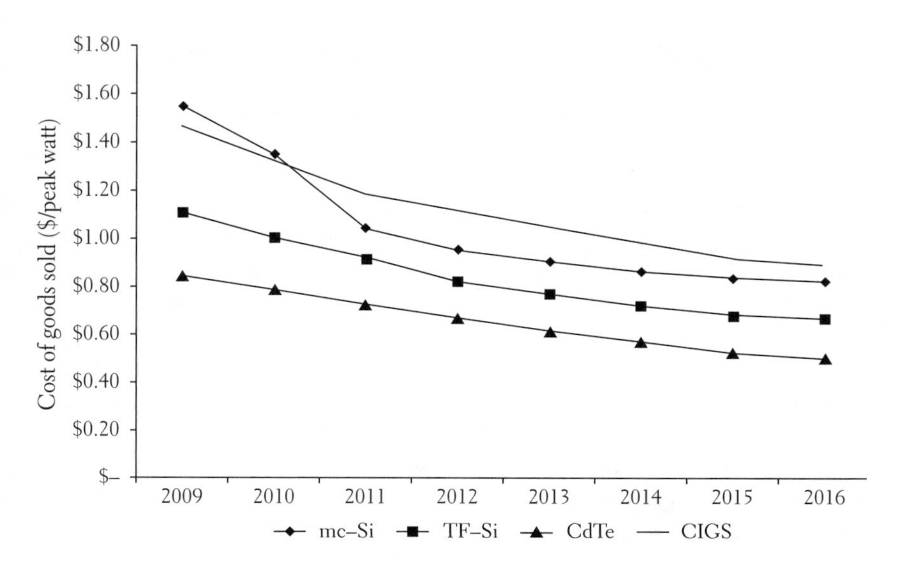

FIGURE 10.1 Cost per peak watt of solar cells
SOURCE: Lux Research, Inc., www.luxreasearch.com, created by Matthew Feinstein.

latter, proprietary equipment and processes are still required, and thus there are much higher barriers to their production. Consistent with the discussion in Chapter 7, vertical disintegration between equipment suppliers and manufacturers has emerged faster for silicon than for other technologies and this has reduced the barriers to entry for manufacturers of silicon solar cells. The end result is that demand-based subsidies are funding the production of the higher-cost silicon-based cells and are probably encouraging firms to produce them because they enable even high-cost producers to turn a profit.[20]

The lower costs of CIGS and CdTe solar cells exist in spite of the larger scale of substrates and equipment used for producing cells made from crystalline silicon. As analyzed in Chapter 3, increasing the scale of substrates and production equipment is an important method of reducing the cost of chemicals, LCDs, and semiconductors. Similar things are now occurring with solar cells, partly because the equipment used to manufacture ICs, LCDs, and solar cells is very similar. The largest difference between these technology paradigms is that reductions in feature size are far more important for ICs than for LCDs and solar cells, and increases in "substrate size" are far more important for LCDs and solar cells than for the wafer size of ICs. For LCDs and solar cells (and for their glass substrates), it is the deposition of thin-film materials on a scale of multi-meter substrates, whereas for semiconductors it is the forming of nanometer patterns on a 0.30-meter-diameter wafer, that drives large cost reductions. Nevertheless, the similarities in processes and equipment and the similar benefits from geometrical scaling enable much of the equipment for producing solar cells to be based on designs that were developed in these other industries.

But how much more can costs be reduced because of further increases in substrate and equipment scale? Since silicon-based solar cells, in particular crystalline cells, are produced with larger substrate sizes than are other cell types (see the first column in Table 10.3), there are probably more opportunities for further reductions with solar cell materials other than silicon. This is further evidence that the cost of producing thin-film solar cells, in particular CdTe or CIGS cells, is potentially lower.

We can use historical data for LCDs to roughly forecast potential cost reductions for solar cells from increases in the scale of substrates and production equipment. Data on reductions in the capital cost of LCD-manufacturing equipment and facilities obtained with increases in substrate size was provided in Chapter 3. If the capital costs of solar cell production facilities exhibit a similar pattern, and if we know the approximate substrate sizes for the cost data points shown in Figure 10.1, we can *roughly* forecast the reductions in capital cost of production from the use of larger substrate sizes. This is shown in Table 10.3. Based on these rough estimates, we can estimate that the 2011 cost per peak watt shown in Figure 10.1 might fall from $1.15 to $0.74 for CIGS, from $1.10 to $0.33 for thin-film amorphous silicon, and from $0.70 to $0.33 for CdTe as larger substrate sizes, particularly 5.7 square meters (Generation

TABLE 10.3

Expected benefits from larger solar cell substrates

Technology	Substrate size (m^2)	Equivalent generation of LCD equipment	Ratio of Generation VIII to current generation (capital costs per area)	Expected cost per peak watt in Generation VIII
Crystalline silicon	5.7	VIII	1.0	NA
Microcrystalline silicon				
CIGS	0.72	V	0.64	$0.74
CdTe	0.72	IV	0.29	$0.20
Amorphous silicon	1.43	IV	0.29	$0.33

SOURCE: Substrate sizes from Signet Solar (2007).

NOTE: LCD, liquid crystal display; CIGS, cadmium indium gallium selenide; CdTe, cadmium telluride.

VIII), are utilized. Larger substrates, which are still occurring with LCDs, could drive these costs even lower.

Although the figures shown in Table 10.3 are clearly very approximate, they suggest that there is a large potential for further reductions in solar cell cost just from increasing the scale of production equipment. If it also becomes possible to double the efficiencies of cells produced in factories, the cost might fall an additional 50 percent. Finally, further reductions in material thickness offer additional opportunities to lower cost. However, for crystalline-based solar cells, the already large substrate sizes, the smaller opportunities for increasing efficiencies, and the difficulties of reducing the thickness of active materials suggest that there is far less potential for reducing costs than there is for thin-film solar cells.

Recent and Future Developments

Research on improving the efficiency of solar cells and reducing material thickness continues. Although improvements in laboratory efficiencies of semiconductor-based solar cells have slowed considerably over the last 20 years, there appears to be large potential for improvements in the efficiency (and costs) of the more recently developed organic and dye-sensitized cells, and efforts are increasing in this area. Organic cells were first produced in 1986 following research on conductive polymers; dye-sensitized cells were first identified in 1991. With efficiencies as low as 2 percent for cells produced in factories (see Table 10.2), significant opportunities for improvements in efficiencies and thus costs remain. Like many of the examples in Chapter 2, these improvements come from finding materials that better exploit the photovoltaic phenomenon. Since scientists have been able to find organic materials with better luminescence per watt and better electron mobility, it is likely that they will be able to find organic materials that absorb more light, and more frequencies of light, and thus have higher photovoltaic efficiencies.

In addition, efforts to increase the efficiencies of solar cells are important for reducing material thickness. Unlike LCDs and semiconductors, a key challenge here is preventing efficiencies from falling along with thicknesses. As active materials become thinner, the time spent by photons in them drops and so do the chances of an individual photon creating an electron-hole pair before it passes through. Researchers are attempting to increase the amount of time a photon spends in the relevant material by "trapping" the light in the material. One method is to deposit materials with textured surfaces that cause light to enter at an angle and thus increase the chance that it will bounce around the material and take a long path through it.

Such efforts are also relevant for multi-junction solar cells. Most implementations of multiple junctions on a single substrate have involved expensive materials such as gallium arsenide and indium phosphide, but more recently researchers have been combining less expensive materials such as amorphous and microcrystalline silicon. Implementing multiple junctions and reducing the thicknesses of the active materials while maintaining or increasing efficiencies will require many of the same types of improvements in materials and processes that were mentioned earlier.

Summary

One theory is that costs are reduced as automated equipment is implemented and organized into flow lines in response to increases in cumulative production. However, this section's analysis of solar cells suggests that the reality is more complex. There are some benefits from implementing and organizing automated equipment, but combining different thin-film layers in a single substrate, further reducing the thickness of materials, further increasing efficiencies, and then increasing the scale of production equipment will also determine the extent to which costs fall. The problem is that demand-based subsidies may not be encouraging firms to do these things. In particular, most global production is for single-crystalline and polycrystalline silicon and even that for thin-film is for amorphous silicon, all of which are much more expensive than CdTe (amorphous silicon's cost are similar to those of CIGS) This suggests that subsidies may be encouraging firms to produce the existing technology rather than make the improvements that were just summarized.

As mentioned, a major reason that silicon solar cells are produced more than others, such as those using CIGS and CdTe, is that it is easier to obtain equipment, processes, and turnkey factories for them. Furthermore, many noted that one of the largest suppliers of solar cell manufacturing equipment, Applied Materials, scaled up its equipment for polycrystalline and thin-film amorphous silicon before reductions in material thicknesses and improvements in efficiency were made. Thus some of its customers are now high-cost producers.[21] Just as Chapter 3 described how effective scaling required improvements in a wide variety of supporting components in many technologies, benefiting from increasing the scale of solar cell production equipment also requires improvements in materials, processes, and material thicknesses.

As with wind turbines, demand-based subsidies are not encouraging the right activities. By encouraging production with these subsidies, we encourage firms to produce the existing technology along with both effective and ineffective increases in the scale of production equipment. We want firms to improve efficiencies and reduce material thicknesses, and we want them to do these things before they begin increasing scale. To encourage this, we should spend less on demand-based subsidies and more on R&D and on supply-based incentives such as government subsidies and R&D tax credits.

ELECTRIC VEHICLES[22]

The primary difference between electric and conventional vehicles is the replacement of an internal combustion engine (ICE) and a gas tank with a motor and a battery (or other energy storage device) that is charged using electricity. Otherwise, the electric vehicle is essentially the same as a conventional vehicle and so other subassemblies can largely be borrowed from conventional vehicle manufacturing. These subassemblies and their materials have already benefited from increases in production scale. The theory that the costs of electric vehicles will fall as more of them are produced is thus very misleading.

The relevant technology paradigm here is not for the vehicle but rather for the battery, the motor, and the battery's source of electricity. For the source of electricity, the question is whether it should be a conventional one such as coal or oil or a clean one such as solar or wind. Since most electricity is still generated by coal and oil, electric vehicles are currently not much cleaner than conventional vehicles, and they will not become "cleaner" until cleaner forms of electricity generation such as wind and solar become cost-effective enough for wide implementation.

For the battery and motor, the key dimensions of performance are energy and power densities. Along with a vehicle's weight, these directly determine range and acceleration. Unfortunately, batteries have about one-fifteenth the energy density of gasoline, and electric motors have slightly lower power densities than those of ICEs; also, the rate of improvements in the energy density of batteries is very low. As discussed in Chapter 2, energy density has been improved by about 10 times in the last 100 years, and some sources show that it took about 15 years (between 1994 and 2009) to double the energy density of lithium-ion batteries, which have the highest storage densities of current batteries. However, lithium-ion batteries are more expensive than lead batteries on a weight or volume basis, and experts do not expect higher rates of improvement in storage densities in the future, even for new lithium-based battery types such as Li-air.[23] Some expect that flywheels, with similar storage densities as batteries, or capacitors, because of their faster rate of improvement in energy storage density (see Figure 2.2) are better candidates than batteries.[24] Others argue that compressed-air engines are a better source of propulsion because they cost much less than either batteries or ICEs.

Of course, the limited energy and power densities of batteries and motors have resulted in an intermediate solution: so-called hybrid vehicles that include both batteries and ICEs. However, because hybrids contain both forms of propulsion, they will always be more expensive and heavier than conventional vehicles. Thus their diffusion will depend on whether higher fuel efficiencies and lower carbon emissions will compensate for their higher prices (and on the extent to which governments subsidize electric vehicles or tax carbon emissions and/or gasoline). Given the public's greater emphasis on upfront price than on life-cycle costs and given the unwillingness of most governments to tax carbon emissions and/or gasoline, the diffusion of hybrid vehicles will probably not occur in large numbers. And even if it does occur, they will have a very small impact on gasoline use. Instead, diffusion of electric vehicles will have to wait for improvements in the energy and power densities of batteries or other energy storage devices, which may take decades.

Like the analyses of wind turbines and solar cells, this brief analysis of electric vehicles partly highlights the problems with the theory that costs fall as cumulative production rises and with the emphasis on demand-based subsidies to stimulate increases in cumulative production. I say partly because lithium-ion batteries for vehicles are about five times as costly as those for electronic products, perhaps because of differences in production volumes. This means that dramatic increases in production volumes might lead to lower costs per kilogram for lithium-ion batteries. Given that 1.2 billion batteries were produced in 2008 for electronic products such as laptops and mobile phones,[25] if ten batteries are needed for each electric vehicle, production volumes for vehicles of about 100 million could lead to costs per kilogram for electric vehicle batteries that are similar to those for electronic product batteries. The obvious problem with this is that subsidizing the production of 100 million vehicles would be in the hundreds of billions of dollars. Another problem is that, even if the costs per kilogram were improved, the energy density shortfalls would make these vehicles much less attractive to buyers.

The alternative is to spend more money on R&D, although here the benefits may be lower than they are for solar cells and wind turbines. One hundred years of data for all batteries (10 times in improvements) and 15 years of data (doubling) for lithium-ion batteries suggest that the battery technology paradigm does not include the kinds of improvements that we need. Instead, it might be better to spend more on other technologies such as flywheels or capacitors, which were mentioned previously, or on radically different batteries such as thin-film types that benefit from geometrical scaling in the same way that solar cells do[26] and nanoscale types that benefit from geometrical scaling at the nanoscale.[27] Also important would be spending on completely different technologies such as compressed-air engines.

In summary, it is hard to be as optimistic about electric vehicles as about solar cells or even wind turbines. Although increases in production volumes as the scale of equipment is increased will probably lead to falling costs, the large increases in volumes needed, the high cost of subsidizing these high volumes, and the slow rate of improvement in battery storage density suggest that we need to search for new and better storage solutions and that supply-based rather than demand-based approaches are probably better at stimulating this search. Furthermore, we need to help students search for new types of solutions, which requires us to give them the tools to do so. Understanding the technology paradigms of various solutions and thus the potential for improvements in these solutions is one important tool.

FUSION POWER

Fusion is how the sun generates its power. In fusion reactions, two light atomic nuclei, such as hydrogen, fuse together to form a heavier nucleus, such as helium; in doing so, they release a comparatively large amount of energy. The challenge is to generate the temperatures found in the sun without destroying the containment vessel. The two ways of doing this are (1) heating a hydrogen pellet with a high-energy laser and(2) containing a high-energy plasma of hydrogen gas within high magnetic fields.

The main measure of progress in either method is in terms of the time in which a fusion reaction is maintained and the power generated during that time. According to Rob Goldstein, former director of Princeton's Plasma Physics Laboratory, scientists have been able to increase the time from 0.01 second to 1 second and the power output from 0.01 watt to 10 million watts (10 megawatts)—an improvement in energy production of ten orders of magnitude over the last 40 years. If we want such a reactor that operates for a year and produces 1000 megawatts of power, we need another nine magnitudes of improvement, which some expect by 2033.[28]

Several conclusions can be drawn from this. First, the rate of improvement is quite rapid—much more rapid than is currently being accomplished with wind turbines, solar cells, or electric batteries. Second, the traditional learning curve tells us nothing about these improvements since we do not yet have any commercial production. Third, although maintaining the rate of improvement will require large amounts of money, these amounts are probably comparable to the subsidies we currently spend on solar, wind, and electric vehicles.[29] Fourth, the cost of the electricity generated from fusion power will probably depend on the relative benefits of scaling, and this should be a topic of research for management and economics scholars.

CONCLUSIONS

This chapter argued that the concept of technological change discussed in this book can help engineers, managers, policy makers and analysts, professors, and even students better analyze the potential for clean energy, and thus devise better energy policies, than can theories about cumulative production driving cost reductions (or about dramatic improvements in performance and price occurring after a discontinuity emerges). It was proposed that through improvements in the efficiency with which basic concepts and their underlying physical phenomena are exploited, as well as through geometrical scaling and improvements in "key" components, we can learn more about the future costs of clean energy technologies. For wind turbines, increasing scale through lighter and stronger materials and finding new designs that benefit from increases in scale are more important than cumulative turbine production in a factory. For solar cells, better and thinner materials and increased scale of substrates and production equipment are more important than cumulative production as well. Furthermore, since the scale of production equipment should be increased only after efficiency and reductions in material thicknesses are increased, these latter improvements are more important than stimulating production. For electric vehicles, an emphasis on cumulative production may make more sense than it does for wind and solar cells. Nevertheless, it is the technology paradigm for energy storage densities that is important, and policies should be based on an understanding of it. Finding materials with higher storage densities is the most important long-term challenge.

This book's portrayal of technological change has important implications for the issue of public support of clean energy and in particular for the issue of demand- versus supply-based approaches. Most current public support focuses on demand-based approaches, such as direct subsidies, feed-in tariffs, renewable portfolio/fuel standards, and systems benefit charges, partly because many believe costs fall as cumulative production rises.[30] This chapter's analysis suggests that we should spend more on R&D through, for example, direct project funding and R&D tax credits, including research on higher strength-to-weight materials for wind turbines, thinner and more efficient solar cells, and higher-density energy storage. Not only will such research have a greater impact on improvements in performance and cost than will demand-based subsidies; it will also, probably, have a greater impact than will the funding of specific companies such as Solyndra because a broad research program supports a broad range of designs for these technologies.

The conclusions here are consistent with Thomas Friedman's observation that, while we need a green revolution, we are having a green party, and it accords with Bill Gates's argument that current demand-based subsidies are really "feel-good" approaches.[31] Both argue that money is being spent very unwisely, and this chapter reaches the same conclusion. However, while management and economics scholars would probably interpret these issues solely in terms of institutional problems in governments, I believe that our models are partly responsible for these problems because our models emphasize that costs fall as cumulative production increases and that increasing cumulative production is the most important thing that governments can do. This book argues that this is an inaccurate recommendation and that management and economics scholars should develop and promote more accurate models of technological change in order to help governments make better decisions.

11

Conclusions

There are many factors that drive the emergence of new industries. They include new and unexpected customers, changes in consumer taste, technologies with a large potential for improvements in cost and performance, technological discontinuities, and vertical disintegration. For many of the industries discussed in this book, multiple factors led to formation. For example, new and unexpected customers played a role in the formation of the mainframe, minicomputer, PC, and handheld computer industries and in the formation of industries that manufacture cassette players, VCRs, CDs, and memory and microprocessor ICs. However, since little research has explored the factors that determine a technology's potential for improvements in cost and performance, or that determine the timing of technological discontinuities, these factors have been the focus of this book. The basic argument is that such an understanding helps us both better choose those technologies that warrant further investigation of business models, strategies, and policies, and better identify the appropriate business models, strategies, and policies.

WHAT DETERMINES THE POTENTIAL OF A NEW TECHNOLOGY?

Unless technologies have a potential for large improvements in cost and performance, they may never offer a superior value proposition to a wide number of users and thus enable an industry to emerge. A major reason that there has been more discontinuous change over the last 50 years in the electronic than in the mechanical sector is that ICs and other electronic components have had a potential for much greater improvements in price and performance than have mechanical components. For example, personal flight, underwater, and space transportation have not emerged as industries largely because there have been few improvements in them or their key components, in spite of large expectations for them since the 1960s.

This book also challenges many of the existing theories about what drives improvements in performance and cost and thus what determines the potential for a new technology. For performance, one theory is that *dramatic* improvements follow a technological discontinuity. For costs, one theory is that the cost of producing a product drops a certain percentage each time cumulative production doubles, as automated manufacturing equipment is introduced and organized into flow lines.[1] Both theories are implicit to some extent in Christensen's theory of disruptive innovation: once a low-end product finds an unexplored niche, an expansion in demand leads to greater investment in R&D and thus improved performance and cost.

This book showed that there are four broad methods of achieving improvements. While cumulative production in a system may indirectly lead to some improvements, it is not their direct driver. First, improving the efficiency by which basic concepts and their underlying phenomenon are exploited was important for a broad number of technologies. The efficiencies of engines and electricity generation were incrementally improved over many years through increasing temperatures, pressures, and scales. For other technologies such as batteries, lighting, display, transistors, and solar cells, improving efficiencies primarily involved finding materials that better exploited a physical phenomenon. Many of these materials could be found before a technology was commercialized; this is now largely the case with OLEDs and organic transistors.

Second, some technologies benefit from the introduction of radical new processes. Both new processes and new materials were particularly important for chemicals, metals, and other basic materials that exist at low levels in an overall "nested hierarchy of subsystems."

Third, some technologies benefit from large increases or large reductions in scale. Increases in scale were particularly important for some forms of production equipment, engines, electricity generation, and transportation equipment, and these increases also enabled increases in the efficiency of engines and electricity generation. Reductions in scale were important for ICs and magnetic and optical storage.

Fourth, the performance and cost of some systems, particularly those at high levels in a nested hierarchy of subsystems, are driven by improvements in "key components." When a key component benefits from geometrical scaling, the system can experience dramatic improvements in performance and cost. For example, improvements in steam engines led to better transportation equipment such as locomotives and steamships. Improvements in internal combustion engines led to better automobiles and aircraft. Most important, improvements in ICs and magnetic recording density led to better computers, televisions, and mobile phones, and a better Internet, as we saw in Parts II and IV.

Consistent with other research, this book also found that these four broad methods of achieving improvements depend on improvements in supporting components and advances in science and that, for this reason, accelerating system improvements is not a simple task. In Part I, we saw that scientific advances formed a base of knowledge that helped scientists and engineers find new materials, radical new processes, and new levels of scale. Part II showed that this base of knowledge helped scientists and engineers find new concepts that formed the basis for technological discontinuities in computers, magnetic recording and playback equipment, and semiconductors (as well as improvements in key components). Chapter 10 demonstrated how improvements in supporting components, such as new materials for wind turbines and increases in efficiency or reductions in material thicknesses for solar cells, are needed before the full benefits from geometrical scaling can be achieved with these technologies. Furthermore, demand-based subsidies may encourage increases in scale before these components and advances become available, and thus supply-side rather than demand-based policies are probably appropriate.

COMPONENTS, SYSTEMS, AND DISCONTINUITIES

Part II explored the timing of technological discontinuities in terms of the time lag between the characterization of concepts and architectures and a discontinuity's commercialization. Its conclusions were that the performance or cost of one or two components was the major reason for this time lag, which was often very long. Exceptions were discussed at the ends of Chapters 6 and 9.

Components

Incremental improvements in components impacted system performance (and the emergence of discontinuities) in several ways. For computers, magnetic recording and playback equipment, and semiconductors, most performance improvements came directly from component improvements. Improvements in ICs have been the major source of better cost, processing speed, response time, and portability of computers. Improvements in magnetic recording density have been the major source of better cost, recording capacity, portability, and quality of audio and video recording and playback equipment. Improvements in semiconductor manufacturing equipment have been the major source of improved defect densities and feature size and, in combination with the benefits from scaling in ICs, have enabled many orders of magnitude increases in the number of transistors per chip.

Some component improvements may even have been more important than the discontinuities themselves. For example, even without the introduction of microprocessors, ASICs, and FPGAs, ICs would still have become cheaper, more powerful, and more widely used, and would have led to improvements in the performance and price of computers, digital switches, mobile phones, and other electronic products. Similarly, even if PCs had not been introduced, computers would have become cheaper and more accessible through connections to mainframe computers with dumb terminals and telephone lines. In fact, even without the development of packet communication, improvements in computers, dumb terminals, and telephone lines, along with those in semiconductor lasers, photodiodes, and the purity of glass, would have enabled users to download and upload information to databases, which might be called the formation of the Internet. In this way we see the benefits of finding technologies with a large potential for improvements.

One reason for emphasizing the importance of individual components is that it helps us understand what drives the performance and emergence of discontinuities in some systems better than does a general concept such as "novel combinations of components." An emphasis on a novel combination of components implies that any component can theoretically be combined with any other component and that improvements come from finding the best and most novel combination. For competition between firms, there is much truth to this argument, in that the most successful suppliers of computers, audio and video recording and playback equipment, semiconductors, mobile phones, and other high-technology products were probably better at combining components than others were. For example, many argue that Apple did this with the iPod, iPhone, and iPad. However, in terms of when these "systems" (including those addressed in Part II) became economically feasible, improvements in individual components probably had a larger impact on timing than did their novel combinations. If we consider the iPod Nano instead of the first iPod, the timing of their feasibility primarily depended on when processing ICs, memory ICs, and displays had reached certain thresholds of performance and price. Part of Apple's success was that it understood when its products might become economically feasible. In other words, not only did specific components largely determine feasibility; it was the same three components that determined the feasibility of these three "systems."

Trade-Offs and Technological Discontinuities

One way to analyze the timing of system-based discontinuities is to look at the trade-offs that both suppliers and users make between a system's price and performance and between the different dimensions of its performance, and then to measure the impact of component improvements on these trade-offs. For both suppliers and users, trade-offs are determined by the technology paradigm and the concepts and principles that it defines. For users, the notion of trade-offs has greater potential explanatory power than does Christensen's notion of "technology overshoot" because it can be applied to any high-end or low-end innovation.

Chapter 5 described at least two examples of technologies that entered from the high end and whose diffusion can be better characterized in terms of changes in trade-offs. Improvements in lasers, photodiodes, and metal-coated plastic disks enabled increases in audio and video quality (and reductions in size) that were much higher than the quality of both magnetic tape and phonograph records. This caused users to change the trade-offs they were making between quality and price. Chapter 6 described many examples of discontinuities that are better characterized in terms of trade-off changes than in terms of technology overshoot. These include the change from bipolar to MOS and CMOS transistors and the emergence of microprocessors, ASICs, and ASSPs.

Many examples were provided that focused on the perspective of suppliers and designers. For example, computer engineers/designers initially employed batch processing to achieve a high utilization of expensive mainframe computers. As the costs and the performance of ICs improved, however, the importance of utilization decreased, and it made more sense for computer designers to provide faster feedback from each keystroke through personal computers. For audio and video recording equipment, complex tape-handling systems were initially necessary to provide acceptable levels of quality. As the recording density of tape improved, however, this trade-off changed and it became possible to use simpler and thus cheaper systems. Improvements in magnetic recording density affected both the general trade-off between quality, cost, and complexity and the more specific trade-offs between tape length, width, and speed. Nevertheless, it is likely that improvements in components other than those associated with magnetic recording density would have had different impacts on the trade-offs inherent to magnetic tape systems. For example, if the speed of tape-handling systems had experienced several orders of improvement, a different set of design changes would have been made.

Similar examples can be provided for semiconductors. Semiconductor engineers made trade-offs between germanium and silicon, between lower and higher levels of integration on a single IC chip, between low and high power consumption, and between different organizations of transistors on chips. In each case, improvements in equipment impacted the trade-offs that suppliers/designers were making, and any changes in the rate of these improvements would have affected the rate at which these discontinuities emerged and diffused. For example, slower improvements in defect density would have delayed or may have even prevented the IC's emergence. Similarly, slower increases in the number of transistors per chip would probably have slowed the diffusion of MOS and CMOS transistors, as well as that of microprocessors, ASICs, and ASSPs.

Finally, the notion of trade-offs can apply in terms of geometrical scaling. When a component has a large impact on the performance and costs of a system and it benefits more from geometrical scaling than the system does, it is likely that system discontinuities will frequently appear. For example, one of the reasons for the many discontinuities in computers and magnetic recording equipment was that these systems did not benefit from increases in scale or, in some cases (e.g., computers), actually became more expensive as scale was increased. If they had benefited from increases in scale, improvements in their components might have merely led to increases in system size and not to the emergence of scaled-down computers such as PCs.

Thresholds of Performance and Price for Components

A simpler way to analyze the timing of system-based discontinuities is to analyze their thresholds of performance and price and those for their key components. Just as minimum performance thresholds and maximum price thresholds exist for new systems that can be defined as technological discontinuities, so thresholds exist for some components as long as one component has a large impact on the system's performance and cost. This was the case for the technologies analyzed in Chapters 4, 5, and 6, which defined the minimum threshold of performance in ICs (in terms of number of transistors per chip) for each discontinuity in computers, in equipment (in terms of minimum IC feature size) for each discontinuity in semiconductors, and in magnetic tape (in terms of recording density) for each discontinuity in magnetic tape–based systems. These chapters focused on minimum thresholds because many of the measures of performance for computers, ICs, magnetic tape, and equipment include aspects of both performance and price. For example, the main trend identified by Moore in his original 1965 *Electronics* article[2] (see Figure 6.1) represented the movement of minimum cost curves over time, where the minimum cost for a specific point in time represented a specific number of transistors per chip. Thus, the analyses of minimum thresholds of performance for computers and ICs in Chapters 4 and 6 include price; similar logic applies to magnetic recording density and thus to the analysis in Chapter 5.

New Applications and Users

One challenge with estimating thresholds of performance or price, or with analyzing changes in trade-offs, is that all applications and users are typically not known in advance.[3] This was certainly the situation with the cases considered in Part II. Although mainframe computers were first purchased by existing users of punch card equipment, subsequent discontinuities such as minicomputers, PCs, and PDAs were first seen in new applications and often by new users. For semiconductors, although defense contractors were the first user of discrete transistors and ICs, the ICs that emerged from other discontinuities were first used in pocket calculators and computer memory (i.e., MOS ICs), digital watches (i.e., CMOS ICs), and calculators and specialized equipment (i.e., microprocessors).

For audio and video recording and playback equipment, the first users of tape-based audio discontinuities changed from radio broadcasters for reel-to reel tape players to automobile owners for 8-track players, and to a mixture of music companies and consumers for the discontinuities associated with digital recording and playback. The first users of video discontinuities changed from television broadcasters for the Quadruplex to education and training for helical scan, to news gathering for the camcorder, and to a mixture of television broadcasters and consumers for discontinuities associated with digital recording and playback. If formal calculations for minimum thresholds of performance or maximum thresholds of price for magnetic tape, semiconductor-manufacturing equipment, and ICs were carried out, these changes in initial applications would impact them. The results are consistent with the notion that firms should look for new applications and new customers, which is a central message of Christensen's theory of disruptive innovation.

Components and Other Industries

Another conclusion from Part II's analysis of technological discontinuities is that many of the improvements in components were driven by other industries. For computers, improvements in vacuum tubes that led to the first mainframes in the 1940s had been primarily driven by the market for radio receivers since the 1920s. Improvements in ICs that led to minicomputers in the mid-1960s were driven by military applications. Improvements in microprocessors that led to PCs were driven by the markets for calculators and aviation and scientific instruments. And improvements in LCDs that led to laptop computers and PDAs were driven by the markets for portable calculators, digital watches, and later camcorders.

For magnetic recording and playback equipment, improvements in coating technology, magnetic coils, and plastics that made the first reel-to-reel tape players possible in the 1940s were driven by the cigarette, telecommunication, and radio industries, respectively. Further improvements in magnetic recording density were partly driven by data applications in the computer industry, as were many of those in microprocessors and ICs. For camcorders, early improvements in CCDs were driven by facsimiles, and improvements in LCDs were initially driven by digital watches and portable calculators.

Although Chapter 6 did not explore this issue to the extent that it was explored in Chapters 4 and 5, much of the earliest semiconductor-manufacturing equipment was borrowed from other industries. For example, vacuum technology, which was necessary for even early semiconductor-manufacturing equipment, was borrowed from the aerospace and nuclear energy industries.[4] The early equipment and processes for photolithography were borrowed from the printing industry, and later improvements were borrowed from the optics industry.

Similar things are now occurring in clean energy. As discussed in Chapter 10, the equipment for solar cell manufacturing is being partly borrowed from the LCD and other industries, just as the equipment for LCDs was partly borrowed from the semiconductor industry. The materials for wind turbines are being borrowed from the aircraft industry, and the batteries for electric vehicles are being borrowed from electronic industries. In all of these cases, we can say that the new industry is "free-riding" to some extent on an existing industry.[5]

The importance of this free riding gives us another reason to see an emphasis on cumulative production in a factory as the main driver of cost reductions, and the resulting emphasis on demand-based incentives, as misleading. Many of the improvements in system cost and performance are not only from activities done outside of a system's factory; many also derive from improvements in components occurring outside the new "system."[6] These improvements in components or advances in science can improve the "potential" cost and performance of a system before it is implemented. For components, it is only as they are modified for a new system that the demand for the new system has an impact on their costs. For example, as microprocessors and other ICs were modified for PCs in response to increases in PC demand, their costs were impacted by that demand. Even after these components were modified, however, reductions in feature size that occurred as a function of Moore's Law probably had a larger impact than PC demand on their rate of improvement.

Second, the emergence of contract manufacturing means that many new systems can benefit from economies of scale in production even before there is large demand for them. Only if special-purpose manufacturing facilities are built for the new system does demand for it have a large impact on manufacturing cost. Third, to understand the potential for cost reductions in a special-purpose production system, it is better to analyze the potential for its geometrical scaling (and scaling in the product design itself) than to just analyze cumulative production and implement policies for stimulating it with demand-based incentives. Fourth, for many electronic products, the cost of implementing and operating special-purpose equipment is trivial compared to the cost of components and the cost of amortizing product development costs (including software) over individual units. As noted in several places in this book, less than 5 percent of an electronic product's cost involves assembly; the rest involves electronic components, many of which are standard.

INNOVATION-RELATED SKILLS

The importance of free riding and the overall importance of "key" components also tell us something about the skills that are needed by individuals, organizations, and society as a whole. One skill involves the introduction of discontinuities. An example is Apple's skill both in combining components in novel ways and in understanding, by analyzing key components, when the iPod, iPhone, and iPad would become economically feasible. The notions of trade-offs and thresholds as methods of better understanding when discontinuities might become economically feasible can be thought of as tools whose use requires a general set of skills and knowledge about various technologies.

Other skills are important for improving a technology's performance and cost through, for example, improving efficiency or implementing geometrical scaling. Such skills are necessary for individuals, organizations, and societies. For efficiency, the ability to find materials that better exploit a physical phenomenon, and an understanding of the extent to which efficiencies might be improved, involves a number of skills at the individual, organizational, and societal levels. The ability to benefit from increases or reductions in the scale of a technology, and an understanding of the extent to which these benefits might emerge, involves a number of skills as well. Transforming improvements into a competitive advantage at both the firm and country level requires still other skills, and these have received the most attention from management scholars; however, they may not be the most important for society. For example, which firms become the leaders in clean energy is certainly less important for our planet than whether we choose, invest in, and implement clean energy technologies that have the most potential to become economically feasible.

A key point here is that different tasks and different technologies require different skills. Thus attempts to identify general skills for implementing and benefiting from innovations will probably not uncover the layers of skills that are needed for individuals, organizations, and societies to succeed and thrive. Some help us understand whether and when technologies might become technically and economically feasible. Others help us improve a technology's cost and performance. Still others help organizations (and countries) transform improvements into a competitive advantage.

Research is needed on all of these skills and on how they enable individuals, organizations, and societies to make improvements and to benefit from them by, for example, creating a competitive advantage. While most current managerial research on innovation uses the number of patents as a surrogate for innovation and the number of papers as a surrogate for advances in science, and applies the notion of learning in factories to learning in laboratories, this book's analysis suggests that we can analyze the skills that influence the rate at which the performance and costs of *real* technologies are improved. We can, for example, measure the rate of innovation in various subfields of biotechnology in terms of improvements that new drugs or bacteria enable in some phenomenon, such as the ability of bacteria to consume CO_2. Not only would this approach eliminate the need to assume that patents are a good surrogate for innovations (when they are not[7]); it could dramatically improve our understanding of innovation and thus the world's ability to more quickly improve the performance and cost of technologies that are needed to solve problems such as global warming.

This chapter now looks at some of these issues in more detail by exploring the implications of this book's analysis for Christensen's theory of disruptive innovations. Christensen's books are widely read by practitioners and scholars; however, his theory emphasizes demand and, in combination with the demand-based theories on technological change that have been assessed here, it can lead to very misleading conclusions. These conclusions illuminate some of the problems with a sole focus on demand. In particular, many followers of Christensen imply or even claim that low-end innovations automatically displace the dominant technology and thus become disruptive innovations.

IMPLICATIONS FOR CHRISTENSEN'S THEORY OF DISRUPTIVE INNOVATION

To refresh the memory of some readers, Christensen differentiates between disruptive and sustaining innovations in terms of an innovation's impact on a system's technological trajectory. Simply put, disruptive innovations disrupt, and sustaining innovations sustain, a system's existing performance trajectory, which is disrupted when it exceeds customer needs and thus leads to so-called technology overshoot and the success of a new trajectory. Although Christensen does not emphasize new concepts or advances in science in *The Innovator's Dilemma*, he does state that both disruptive and sustaining innovations can be either radical (based on both a new concept and a new architecture) or architectural (just a new architecture), as defined by him[8] and others. He recommends that firms, in particular new entrants, focus less on sustaining innovations and more on disruptive innovations, which incumbents have a tendency to overlook.

While some question whether incumbents actually overlook disruptive innovations,[9] I believe there are even larger problems than this with the misinterpretations and conclusions of some followers of Christensen when they use his theory to predict disruptive innovations. Following a description and analysis of these misinterpretations and conclusions, this section describes a better way for firms to analyze any type of discontinuity, whether it is a low-end or high-end one. First, for some, Christensen's classic figure in the *Innovator's Dilemma* (fig. 1.7) implies that improvements in low-end innovations easily occur and enable these low-end innovations to easily replace an existing technology.[10] Such a misinterpretation is supported by the learning curve's emphasis on cost reductions from cumulative production in a final product's factory, the theory that dramatic improvements in performance occur following the emergence of a technological discontinuity, and management failure as the main reasons for a delay between the characterization of concepts and architectures and their commercialization.

Second, the chances of making such a misinterpretation are increased when disruptive innovations are defined as low-end innovations (and radical innovations as high-end innovations), as some do, in spite of the fact that Christensen does not. While all disruptive innovations may be low-end, all low-end innovations *do not* become disruptive innovations. A good definition of a phenomenon focuses on what distinguishes it from other cases, and a disruptive innovation is not just a low-end innovation. In combination with the misunderstandings previously mentioned, defining a disruptive innovation as low-end increases the chance that someone will conclude that low-end innovations naturally become disruptive innovations and thus firms should target the low-end of the market. Not only is there no evidence for the superiority of a low-end strategy; such a strategy can cause students to celebrate low-end products and inferiority, as if customers actually wanted an inferior product.

A third factor contributes to these misunderstandings. If changes in the concepts that form the basis of a product are ignored by practitioners or by students (perhaps because they are not covered in a course), chances increase that students or practitioners will assume that any type of low-end innovation will replace the dominant product. This is often the case when a disruptive innovation is defined as a low-end innovation because its opposite, a high-end innovation, is then defined as a radical innovation; thus innovations based on new concepts are not even considered. In these cases, some will then interpret any low-end product (e.g., any cheap mobile phone) as an innovation and thus as a disruptive innovation that will by definition replace the existing technology.[11]

An equally disturbing conclusion from these interpretations is that the key to disruptive innovations is introducing low-end products such as those that have fewer features or are smaller, as is argued in some research.[12] If we do not understand the importance of new concepts and thus believe that low-end products can be defined as low-end innovations, and if we also believe that low-end innovations automatically become disruptive innovations once demand emerges, it is easy to believe that the key to a disruptive innovation is the introduction of a low-end product. Once you have entered a low-end market with a low-end product, such as one with fewer features or smaller sizes, production volumes increase, improvements occur, and *voila*, we have a disruptive innovation! There are obviously problems with this logic. With respect to smaller sizes, how many people believe that *smaller* automobiles, roads, elevators, buildings, furniture, toiletries, packaged food, production equipment, and clothing will become disruptive innovations?

A final disturbing story about disruptive innovations concerns what can be called a double disruption. Some people argue that a combination of two low-end innovations is even better than one alone. One argument is that producers of solar cells should focus on cells with poor efficiency (e.g., based on photosensitive dyes) because they are inferior to high-efficiency cells, and that they should focus on low-end applications such as electric bicycles. Electric bicycles are defined by some as a low-end innovation because they have lower performance and costs than do the mainstream technologies, automobiles and motorcycles. By selling their low-end solar cells to users of electric bicycles, these solar cell producers will somehow benefit from a double disruption when electric bicycles replace automobiles and the low-efficiency cells replace the high-efficiency cells. But why should these producers care whether their cells provide electricity for lights, mobile phones, or electric bicycles?

This book's analysis leads to a different interpretation of Christensen's theory of disruptive innovation, one that increases the likelihood that practitioners, students, and others will accurately identify potentially disruptive technologies. First, firms must introduce a superior value proposition for the appropriate market. Note that all innovations, whether sustaining or disruptive, must offer a superior value proposition to some users in order to be adopted, and thus there is no such thing as a successful "inferior technology." Instead, the question of whether to enter the low end or the high end of a market depends on the appropriateness of the market for the technology. Some technologies are more appropriate for high-end markets; some, for low-end markets. The first mainframes, laptops, and PDAs were sold to high-end users, as were many other technologies. This includes the first discrete transistors and ICs (for the military), video recorders and players (for broadcasting companies), electronic calculators (for engineers), digital cameras (for high-income consumers), camcorders (for electronic news-gathering organizations), digital watches (for high-income consumers), LEDs (for consumers of expensive digital watches and now expensive lighting systems), LCDs (for use in in cockpits, calculators and digital watches, laptops, and expensive TVs), optical discs such as CDs and DVDs (for high-income consumers), mobile phones (for high-end consumers), semiconductor lasers (for the military), packet systems (for universities,) and solar cells (for military and telecommunication satellites). Christensen's key insight was that *some* technologies are more appropriate for low-end markets. For example, most agree that the first minicomputers and PCs, transistor radios and televisions, VCRs, mobile Internet applications, and SaaS and other utility/cloud computing offerings were sold to relatively low-end users as compared to the lead users of the previous technology.

A second and related question is this: In what situations do low-end innovations displace existing technologies and thus become disruptive innovations? As noted in Part II, the rate at which a discontinuity, whether low- or high-end, diffuses across different market segments depends on the rate at which it experiences improvements in performance and/or price and the degree of preference overlap across these segments. Part I argued that the potential for the former can be analyzed in terms of the system's technology paradigm, which includes geometrical scaling. Thus, while the success of a low-end product may encourage firms to invest in such products, thereby increasing the chance that the full potential of their technology paradigm will be realized, without a technology paradigm that enables the relevant improvements, the investments are meaningless. Chapter 3 described how geometrical scaling existed in either a system or a key component in it for most of the disruptive innovations discussed in Christensen's books and papers. An exception is the mini-mill, which is a little more complex. Mini-mills diffused from low- to high-end applications because they were based on a different concept (electric arc furnace) from that behind integrated mills, and this concept benefited from falling electricity costs, the increasing availability of scrap, and the increasing scale of production equipment (from which integrated steel mills also benefit).

However, even if a low-end innovation experiences large improvements in performance and price, it may not diffuse and become disruptive unless it experiences more improvements than does the dominant product. For example, low-end transistor radios and televisions replaced the dominant products because transistors and ICs were initially more appropriate for them and improvements in these transistor and ICs enabled their cost and performance to outstrip the cost and performance of the dominant products, which were using vacuum tubes. Low-end magnetic disks replaced magnetic cores and drums because improvements in magnetic recording density impacted their performance and costs more than they did the cost and performance of magnetic cores and drums. Improvements in the Internet have a much larger impact on the performance of SaaS, which started with low-end users, than on the performance of packaged software. Thus they may cause SaaS to replace packaged software. Similar arguments can be made for other aspects of utility computing and some Web 2.0 applications.

In some cases, the issue is whether improvements in components affect a dimension of performance for which technology overshoot may or may not occur. For example, increases in the recording density of platters caused hard disks to overshoot the needs of most users in terms of drive capacity, which facilitated the replacement of large hard disks with smaller ones. Similarly, improvements in the magnetic recording density of tape caused high-end systems to overshoot the needs of most users, facilitating their replacement with low-end systems such as those with simpler tape handling. Increases in the number of transistors per chip caused mainframe computers to overshoot the needs of most users in terms of processing speeds, facilitating their partial replacement with minicomputers and PCs.

Mobile phones represent a case in which improvements in electronic components such as ICs and LCDs will probably not lead to technology overshoot for many years. Instead, these improvements and those in other components are enabling better applications and user interfaces for *high-end* smart phones and thus are driving their diffusion throughout the market. For this reason, it is highly unlikely that a low-end mobile phone, unless it is based on a completely new concept such as wireless LAN, will form the basis of a new product that displaces current mobile phones. It is important to recognize that the key issue here is not whether a mobile phone is low-end or not but whether it is based on a concept with a different and perhaps superior technology paradigm and whether this superior paradigm will enable the technology to offer a superior value proposition to some set of users. This example reinforces the argument made earlier about the danger of calling a disruptive innovation a low-end innovation (and a radical innovation, a high-end innovation). If managers and students talk and hear only about "high-end" and "low-end" and not about changes in concepts and architectures, they will not understand the key question: Does the low-end innovation involve a change in concept or architecture that will enable it to have a superior technology paradigm?

Finally, one might ask which has a larger impact on the displacement of an existing technology by a disruptive technology: improvement in system performance or the degree of preference overlap. In addition to the importance of geometrical scaling in most of Christensen's disruptive innovations, the results from Chapters 4 and 5 suggest that improvements in system performance, which were driven by improvements in components, sometimes had a much larger impact than preference overlap on the replacement of high-end with low-end innovations. In all cases described as disruptive innovations in these chapters, there initially appeared to be very little preference overlap. Instead, exponential improvements in processing speed for computers and the recording density of tape caused low-end products to displace the existing technologies and thus become disruptive. For example, accounts of early PCs suggests that no one considered the preference overlap between PCs, minicomputers, and mainframes to be large, but PCs displaced minicomputers and to some extent mainframes because the improvements in ICs and thus processing speeds were so large.

Understanding this technological change—rates of improvements in components and systems, the impact of component improvements on system performance and cost, and users' trade-offs between price and various dimensions of performance—is essential for managers in high-technology sectors. While certain skills are needed for this and there is a cognitive aspect to them, it was noted in a previous section that we cannot identify these skills without considering the different ways in which discontinuities emerge, performance and costs are improved, and competitive advantage from these discontinuities and improvements is gained. Unfortunately, few if any management books explicitly recognize this. Even Christensen partly falls into this trap in his most recent book (with Dyer and Gregerson), *The Innovator's DNA: The Five*

Skills of Disruptive Innovation. By ignoring rates of improvement in price and performance, which he had addressed to some extent in previous publications, and solely emphasizing the managerial skills associated with creating low-end innovations, he and his co-authors imply that disruptive innovations are just as likely to occur in technologies that are not experiencing rapid rates of improvements as they are in technologies that are, and that there are skills that are common to all sectors. Clearly, different skills are needed for finding low-end innovations and transforming them into disruptive innovations, and identifying those for which this might be possible.

IMPLICATIONS FOR GOVERNMENT POLICY

The first implication for government policy is a reaffirmation of support for science and technology. It is well established that advances in technology and science have contributed to economic growth, that firms underinvest in both because they cannot appropriate all the benefits from their advances, and thus that governments should financially support them. The analysis in this book showed that both science and technology have played an important role in the formation of new industries. For example, finding materials that better exploit a physical phenomenon was a direct result of advances in technology and, in many cases, these advances were facilitated by a strong knowledge base in the relevant science. Even improvements that involved new processes or geometrical scaling depended on advances in technology, without which there would have been a slower rate of improvement in performance and cost from scaling.

Chapter 10 showed the practical implications of these advances. While some theories suggest that advances in science and technology will naturally emerge as demand for clean energy grows, it was demonstrated that most investment is flowing toward existing technologies and not to advances in science and technology that could be more directly achieved with supply-side policies such as R&D tax credits or direct funding of university R&D. By supporting a broad range of designs, such supply-side subsidies would have a higher chance of success than investments in specific companies.

A second implication for government policy is a reaffirmation of the importance of general-purpose technologies (GPTs).[13] Improvements in some components have played a much more important role in global economic growth than improvements in others; these "components," which might be called GPTs, deserve attention when governments consider R&D spending and subsidies. Identifying and supporting these components is also critical in addressing global challenges such as global warming and other environmental problems, and universities should help students understand the role of these technologies in such challenges.

A third implication for government policy is that identifying technologies that have a large potential for improvements is more important for governments and societies than introducing the right policies for a given technology. Because the world has experienced such rapid changes over the last 50 years, many people believe that achieving improvements in performance and cost is trivial as long as a country or a firm has the will to do so. A good example of this can be found in clean energy, where many believe that improvements have not emerged because of a conspiracy by automobile and oil companies against them. Such conspiracy theorists might also think that if a U.S. president were to make clean energy a priority, just as John Kennedy made reaching the moon a priority in the 1960s, improvements in clean energy technologies would quickly emerge. This interpretation of America's success at putting a "man on the moon" ignores the technological improvements in jet engines and electronics that were made in the 1950s and 1960s for other applications, which made a trip to the moon technically feasible by 1969.

Another example of how many observers overlook the fact that that some technologies have a larger potential for improvements than do others is the idea that we have exhausted the "low-hanging economic fruit."[14] This is in spite of the fact that humankind is experiencing the largest advances in technology, science, and economic growth in its history. Business has always been about grabbing low-hanging economic fruit, and identifying technologies with a large potential for improvements can help us do this. Furthermore, as this book has shown, there is still a lot of fruit just waiting to be grabbed, but it will require that we create the tools and institutions that can help governments, universities, firms, and students find the technologies and achieve the advances in science that make the fruit grow.

A fourth implication for government policy is that the concept of a technology paradigm can help identify and understand technologies that have the potential for large improvements. Simple technology paradigms were described in Chapter 2, and more detailed ones were described in the chapters in Part II. More detailed examples can be characterized if needed. One, mentioned in Chapter 6, is the International Technology Roadmap for Semiconductors. For some time, the semiconductor industry has been developing a road map that describes the improvements in the "components," such as equipment and other related technologies, needed to continue reducing feature size and increasing transistors on a chip. To some extent, other industries such as MEMS, bioelectronics, nanotechnology, solar, and wind have and are developing similar road maps that can be used to prioritize research and development at the firm, university, and global levels. Such road maps can also help us better understand the skills needed to promote development of these and other industries in specific regions. In addition to access to cutting-edge research, firms base their location decisions on the availability of specific skills (e.g., operating photolithographic equipment, scanning electron microscopes, scanning tunneling microscopes and other advanced scientific instruments), so understanding these skills is essential for regions to succeed in attracting new industries.

IMPLICATIONS FOR UNIVERSITIES

Finally, there are implications for universities, specifically in how they train students. As mentioned in Chapter 1, students have an important stake in our future, they have time and energy to think about it, and they are often more open-minded about it than we are. The last point is important because many working people have a stake in a particular way of doing things and thus are naturally more concerned with their own jobs than in creating new industries that benefit large numbers of people. This is certainly a major challenge with clean energy, where, for example, farmers are focused on biofuels and many solar cell specialists are focused on silicon.

This book argues that we can help students think more effectively about the future by providing them with a better background, through a variety of courses, in technological change and the reasons that some technologies experience more improvements than others. One example of such a course is the history of technology, which is usually offered to freshman or sophomore students and usually focuses on the interaction between technological and social change in the 18th and 19th centuries. While this is important, it is even more important for students to learn about the technological change that occurred in the latter half of the 20th century and is still occurring now. The ideas in this book can help students understand the reasons that some technologies replaced others and thus assess and compare the potential of existing and new technologies—something that neither social science nor engineering programs currently do. While engineering programs rightfully focus on detailed design, many observers argue that they should focus more on problem solving than on mathematics.[15] This book argues they should also help students understand technology change: what drives it and why some technologies experience more improvements than others.

Another possibility is a course on analyzing the future. This course would be more appropriate for graduate or advanced undergraduate students and ideally would build on a course on the history of technology. I use this book's concepts in a graduate course to help students analyze a broad range of technologies that have been proposed by leading scientists, such as Michio Kaku in the *Physics of the Future*, Peter Diamandis and Steven Kotler in *Abundance*, and Mark Stevenson in *An Optimist's Tour of the Future*.[16] The analysis includes rates of improvement; possible timing of technical and economic feasibility; and technological, policy, and business bottlenecks to technology implementation. Providing students with a background on a wide range of technologies and requiring them to address specific technologies in their presentations will enable them to study much more interesting technologies then they could in most courses. For example, helping students (and firms and governments) understand how reductions in the features sizes of ICs, including bioelectronic ICs and MEMS (microelectronic mechanical systems), enabled (and continue to enable) the emergence of new electronic systems can lead them to search for new opportunities (i.e., low-hanging economic fruit) in a wide variety of electronic systems. In my recent courses, students used this information to analyze specific types of 3D holograms, 3D displays, MEMS for ink-jet printing and pico-projectors, 3D printing, solar cells, wind turbines, cognitive radio, wave energy, membranes, pico-projectors, wireless charging, and human-computer interfaces.[17] These presentations formed the basis for some of the analyses in Chapter 9.

In business, economics, social science, and even engineering programs, it is also possible to help students analyze new systems and either business models or policies for them. Although many entrepreneurship courses essentially do this, providing the students with the concepts in this book and with a background in specific technologies will enable them to address much more interesting technologies and in far more detail then they currently can. Consider clean energy. By providing students with a background in specific clean energy technology paradigms, including geometrical scaling and the interaction between systems and components, they can go beyond simple topics such as "solar in general," simple value propositions such as "fewer carbon emissions," and simple drivers such as "rising oil prices" (which seem to fall as often as they rise). Furthermore, such courses can build from previous years as presentations from last year's class can help this year's students better do their own. Over time an increasing number of presentations will lead to better courses for future students and thus better presentations on viable new businesses or government policies.

One way to offer such courses is through a global network of professors and students who are addressing the future of complex technological systems. Of course, the Internet makes this much easier than in the past because it makes sharing information easy. The key is to find ways to organize information so that students do not become lost in the details. One twist on this idea is to focus part of such a global network on solving problems in developing countries, in particular rural areas. Many developing countries require solutions that are much cheaper and thus much different from those used in developed countries, but they still need help from developed countries to achieve them. Ideally students will work together to create local businesses that use some core ideas or technologies from developed countries but use local materials and local labor to implement them.

FINAL WORDS

In spite of the growing importance of technology as a tool to solve global problems, most of our understanding is severely limited regarding which technologies are undergoing improvements, which might undergo them in the future, and why some undergo more improvements than others. Simply put, how can we solve global problems when most policy makers, managers, and professors do not understand them enough to choose the technologies that warrant further attention and that can help us find solutions? The media exacerbate these problems. In spite of the many discussions of technology on news programs and documentaries, little data on performance and cost are provided, other than a few mentions of Moore's Law. How about a discussion of trends in battery storage density and solar efficiencies, or of the cost of these and other technologies when clean energy is being discussed? Can we as workers, voters, or students make good decisions about clean energy without this kind of data?

As recipients of public money, universities have a special responsibility to prepare students for the world, for the technological change occurring, and for the global problems they will need to solve. Since improvements in technology are one of the most important tools for addressing these problems, universities need to help students better understand the trends in the performance and cost of new technologies, the reasons for the differences, and the drivers of change. Illuminating these issues was a major goal of this book and continues to be a major goal of my research and courses.

Reference Matter

Appendix: Research Methodology

This book grew out of my interest in how and why the mobile Internet was emerging in the early 2000s (and not earlier) and why in Japan faster than elsewhere. From that interest, I began to look for common patterns of industry formation and reasons for its timing and location. Theories that influenced early ideas for research on this book included evolutionary models of technological change (Nelson and Winter 1982), technological discontinuities (Abernathy and Utterback 1978; Tushman and Anderson 1986; Anderson and Tushman 1990; Utterback 1994; Christensen 1997), complex systems (Tushman and Rosenkopf 1992; Tushman and Murmann 1998; Hobday 1998), network effects, standards, critical mass (Katz and Shapiro 1994; Grindley 1995; Shapiro and Varian 1999; Rohlfs 1974, 2001), and R&D-related capabilities (Vernon 1971; Chandler 1994). Much later, I realized that technology paradigms (Dosi 1982) and geometrical scaling (Sahal 1985; Lipsey, Carlaw, and Bekar 2005; Winter 2008) were central to the issue of industry formation and evolution.

For each chapter, more detailed sources were accessed and the case study method (Eisenhardt 1989) was used to identify common patterns of industry formation and reasons for its timing and location. Consistent with Eisenhardt (1989), specific industries (i.e., cases) from the mechanical, electrical, and electronics sectors were selected. These selections focused on some of the most important industries formed in the 20th century and to a lesser extent the19th century, and on those for which information was readily available.

For Chapter 2, various histories of technology (Cardwell 1995; Crump 2002; McClellan and Dorn 2006) provided a basic background. Also, general web sites such as Wikipedia, Answers.com, and howstuffworks.com; technology-specific web sites; and technical reports were studied to gather more detailed information about specific technologies. Finally, using Eisenhardt's (1989) notion of a cross-pattern search, I compared and contrasted technology paradigms to identify common aspects and to gather data about the four elements of a technology paradigm described in the chapter and the four methods of achieving advances along technological trajectories.

For Chapter 3, I first searched for examples of geometrical scaling in some of the few papers that use this term or related terms such as *scaling* and *increasing returns to scale* (Dosi 1982; Nelson and Winter 1982; Sahal 1985; Freeman and Louçã 2002; Winter 2008). Whenever a relevant discussion of geometrical scaling was found, the original reference was searched using Google and Google Books for details, and in many cases the reference was obtained and read. Specific technologies that were found in these examples (such as electricity, energy, aircraft, furnaces, and engines) were also "Googled" along with terms such as *scale*, *increasing*, and *large* to find other references that might apply. For each instance of geometrical scaling, I identified its type and gathered data on the changes in scale and on cost/price for various levels of scale. For each instance of geometrical scaling, I investigated the improvements in components and the advances in science that were needed to benefit from it. Information on the necessary advances in science was also taken from the same sources used as a background for Chapter 2 (Cardwell 1995; Crump 2002; McClellan and Dorn 2006).

For Part II, I relied on a more formal methodology than for previous chapters. I used primary and secondary literature on the computer, audio recording and playback, semiconductor, and other industries, including academic papers and books from the management, economics, and history fields; practitioner-oriented accounts; and encyclopedic histories to define and classify technological discontinuities using Henderson and Clark's (1990) framework. Although determining whether there was a change in core concept or in architecture when comparing new and previous products is highly problematic, the literature on these industries essentially defines and classified these discontinuities. For example, every history of the computer industry distinguishes between mainframes (along with their predecessor, punch card systems), minicomputers, PCs, workstations, and handheld computers, and the discontinuities in the components used in them, such as transistors, ICs, and microprocessors.

Thus, for Chapters 4, 5, and 6 (Part II), the discontinuities were first classified using Henderson and Clark's (1990) framework. The dates were compared for when the discontinuities first emerged and when the concepts or architectures that formed their basis were identified. Quantitative data on performance and price was collected for most of the discontinuities in order to define minimum thresholds of performance and plot innovation frontiers for computers, magnetic recording and playback equipment, and semiconductors. I carried out a less formal analysis for Chapter 9.

For computers (Chapter 4), building from Ein-Dor's (1985) and Smith's (1988) analysis of diseconomies of scale in computers, data on processing speed and price was gathered for four time periods (1954, 1965, 1990, and 2002) from a web site created by a computer science professor (McCallum 2008). Because there was little data there on personal computers (PCs) between the late 1970s and early 1990s, I supplemented it with data from an academic paper by another computer science professor (Ein-Dor 1985). In total, 75 data points were plotted, and a spreadsheet package was used to choose best-fit curves for the data in each of five time periods, for which each curve represented an innovation frontier. Data on the rates of improvement for some of "key components" was also collected. Data on increases in the number of transistors per chip, better known as Moore's Law, was used to define the minimum threshold of performance needed in electronic components to enable the discontinuities in computer systems. One reason for focusing on the performance of ICs and not on price/cost is that Moore's Law also includes the price/cost of ICs in its determination of the "optimal" number of transistors per chip (Moore 1965).

For magnetic recording and playback equipment (Chapter 5), both qualitative and quantitative data on quality, size, price, and other factors was collected and analyzed, which is described more in Funk (2009a). Since quantitative data on quality was not available, the literature on this industry (e.g., Daniel, Mee, and Clark 1999) was used to estimate relative levels of quality for the various systems. Quantitative data on the data requirements for each major application (analog audio, analog video, digital audio and digital video) was collected along with that on price (inflation adjusted), size, tape width, and tape speed for most tape-based systems. Because the differences in size and price between discontinuities were very large (four orders of magnitude over 50 years) data was only collected for selected years. It was used to plot the evolution of the equipment in terms of quality, size, and price and to show an innovation frontier for tape-based equipment. The data on size, tape width, and tape speed was used to analyze other design trade-offs and to look at how improvements in the magnetic recording density of tape enabled simpler recording systems, including those with thinner and shorter tape and slower tape speeds.

For the semiconductor industry (Chapter 6), I used Moore's original 1965 figure in *Electronics* (Moore 1965), which depicted the innovation frontier for the early years of the semiconductor industry, focusing on the number of transistors per chip. Various reports by the International Technology Roadmap for Semiconductors (ITRS) between 2000 and 2009, web sites (ICKnowledge 2009; Wikipedia 2009c), and reference books (Tilton 1971; Swann 1986) provided historical data on other dimensions of semiconductor performance (e.g., speed, power consumption, heat dissipation, functionality). This data was used to construct innovation frontiers for the later years in the industry (only one example is shown in the chapter). Furthermore, quantitative data on performance was collected for the most important component, semiconductor-manufacturing equipment, in order to analyze the relationship between improvements in equipment and discontinuities in semiconductors, and in particular to define thresholds of minimum performance for equipment in each semiconductor discontinuity. Since Moore's Law also considers price (Moore 1965), Chapter 6 focused on the minimum threshold of performance (and ignored price) for equipment in terms of defect density, minimum feature size, and transistors per chip. Improvements in these and other dimensions of performance showed how the trade-offs that producers and users were making changed when they considered the discontinuities in semiconductors.

Chapter 7 (Part III) used some of the same sources mentioned for Chapters 4 and 6 to define when open standards and thus vertical disintegration emerged. It drew on a previously published paper (Funk 2012b). More specifically, the analysis of the semiconductor sector contrasted the composition of the 48 firms from the 1950s to the mid-1960s examined in Chapter 4 of Tilton's (1971) seminal book with that of 130 firms detailed in Integrated Circuit Engineering's (ICE's) database for 1995. It also analyzed the composition of the largest U.S. semiconductor firms in 1955, in 1965 (Tilton 1971; Braun and Macdonald 1982), in 1983/1984 (UN 1986; Yoffie 1993), in 1995 (ICE 1996), and in 2005 (Edwards 2006). More details are provided in Funk (2012b).

Chapter 8 (Part III) also built from a previously published paper (Funk 2010), which describes the methodology in more detail than was done here. In particular, the chapter used the many books on the formation of specific industries and Eisenhardt's (1989) case study method to summarize and contrast the challenges involved with industry formation.

Part IV relied on a number of reports, analyses, presentations by students, conversations with experts, and a lifetime interest in energy that began in the 1970s. For Chapter 9, I benefited from conversations with Karthik Nandakumar of AStar for human-computer interfaces; Aaron Danner of NUS and Rajan Walia of AStar for lighting/displays; Pei Sin Ng of Rockwell Automation for displays; Han Vanholder of ST-Ericsson for wireless; and Calvin Xu of NUS, Jo Yew of AStar, Alvin Woo Shian Loong of New-Media Express, Justin Lee of Point Star, and Darryl Chantry of Microsoft for Internet content and applications. For solar cells, I benefited from conversations with Michael Gratzel of EPFL; Armin Eberle, Charanjit Singh Bhatia, and Benjamin Sovacool of NUS; Rob Steeman and Ole Sordalen of REC; Christopher Inglin of Phoenix Solar; and Ian Latchford and Hari Ponnekanti of Applied Materials. For wind turbines, I benefited from research done by Srikanth Narasimalu, who was an employee of Vestas for many years. For electric vehicles and batteries, I benefited from many conversations with Palani Balaya of NUS. In many cases, these conversations supplemented presentations by these experts.

Notes

CHAPTER 1

1. The U.S. government expects to spend $150 billion between 2009 and 2019 on clean energy, of which less than $5 billion is expected to involve research on and development of solar cells and wind turbines. This information is from a presentation given by Dan Arvizu at the National University of Singapore on November 3, 2010, entitled "Moving Toward a Clean Energy Future." http://www.nrel.gov/director/presentations_speeches.html. Germany has spent $130 billion on subsidies, which enables it to now receive 0.3 percent of its energy from the sun (Lomborg 2012).

2. Analyses of costs using cumulative production can be found for a variety of industries in Arrow (1962), Ayres (1992), Huber (1991), Argote and Epple (1990), and March (1991). For clean energy, these analyses can be found in Nemet (2006, 2009). The notion that cumulative production is the primary driver of cost reductions is also implicit to some extent in theories of technological change (Abernathy and Utterback 1978; Utterback 1994; Christensen 1997; Adner and Levinthal 2001). For example, Utterback (1994) and Adner and Levinthal (2001) focus on cost reductions through improvements in processes where the locus of innovation changes from products to processes and thus the locus of competition changes from performance to cost following the emergence of a dominant design. Although Christensen's (1997) theory primarily focuses on the reasons for incumbent failure, it also emphasizes demand—how it drives learning and how it leads to improvements in both cost and performance. More specifically, once a product finds an unexplored niche, an expansion in demand leads to greater investment in R&D and thus improved performance and cost for the low-end product. Therefore, the key to achieving improvements in performance and costs is to find these unexplored niches. An exception to these examples can be found in Nordhaus (2009).

3. One analysis of solar cells makes this point (Nemet 2006).

4. See Jason Pontin's interview of Bill Gates, in which the cofounder of Microsoft talks about energy, philanthropy, and management style (Pontin 2010). See also Ball (2010).

5. Schmookler (1966).

6. For example, less than 5 percent of the iPhone 3GS's manufacturing cost in 2009 consisted of assembly costs; the majority of the component costs were standard ICs and depended more on advances in Moore's Law than on cumulative production of the iPhone. See http://gigacom.com/apple/iphone-3gs-hardware-cost-breakdown/.

7. Increases in the number of transistors per chip, better known as Moore's Law, are always presented as a function of time, not cumulative production.

8. Kurzweil (2005).

9. This book's distinction between science and technology and the linear model of innovation is roughly consistent with Arthur's (2007, 2009) characterization. Arthur distinguishes (1) an understanding of a scientific phenomenon; (2) the definition of a concept or principle; and (3) solution of problems and subproblems in a recursive manner. For a broader discussion of the linear model, see Balconi, Brusoni, and Orsenigo (2010).

10. See Albright (2002).

11. Exceptions include Rosenberg (1963, 1969), Freeman and Soete (1997), Mowery and Rosenberg (1998), and Freeman and Louçã (2002).

12. For example, the timing of the industrial revolution differed by decades if not centuries among countries, even those in Europe, and the timing of the formation of the initial banking, insurance, and finance industries may have differed by even larger time spans.

13. For example, Klepper (2007, 2010).

14. The mobile Internet is explored in more detail in Funk (2004, 2006, 2007a, 2007b).

15. Dosi's technology paradigm builds on Thomas Kuhn's (1970) notion of a paradigm shift.

16. While there are many good descriptions of technology change—see, for example, Arthur's (2009) and Constant's (1980) descriptions of jet engines and Hughes' (1983) description of electricity—the potential for improvements in competing technologies is rarely addressed.

17. Many terms are used by scholars. Nelson and Winter (1982) initially used *economies of scale* but Winter (2008) later used *scaling heuristics*. Sahal (1985) uses *scaling*, while Lipsey, Carlaw, and Bekar (2005) use both *geometrical scaling* and *increasing returns to scale*. For organisms, see Schmidt-Nielsen (1984).

18. See, for example, Simon (1962), Alexander (1964), Tushman and Rosenkopf (1992), Tushman and Murmann (1998), and Malerba et al. (1999).

19. For example, Abernathy and Clark (1985), Tushman and Anderson (1986), Utterback (1994), and Henderson and Clark (1990). Technological discontinuities are defined in terms of differences from previous products, but dominant designs are defined by the degree of similarity among existing products (e.g., architectures and components) (Murmann and Frenken 2006). Technological discontinuities and new industries can be thought of as the second stage of Schumpeter's three-stage process of industry formation: (1) invention; (2) innovation; and (3) diffusion.

20. Many scholars have emphasized directions of advance; these include Rosenberg (1969),who used the term *focusing devices*, Sahal (1985), and Vincenti (1994).

21. In order, these elements are similar to Dosi's emphasis on a "specific body of understanding," a "definition of the relevant problems to be addressed and the patterns of enquiry in order to address them," a "specific body of practice," and "the operative constraints on prevailing best practices and the problem-solving heuristics deemed promising for pushing back those constraints" (Dosi and Nelson 2010).

22. Foster (1986) focused on S-curves, and Tushman and Anderson (1986) focused on incremental change punctuated by technological breakthroughs/discontinuities that provide sharp price-performance improvements over existing technologies. Tushman and Romanelli (1985) reinterpreted this phenomenon as *punctuated equilibrium*, a term they borrowed from the field of biology. In biology the theory of punctuated equilibrium says that most sexually reproducing species exhibit little evolutionary change except in rapid and localized cases (Gould and Eldredge 1977). Tushman and Anderson (1986) concluded that technologies also undergo dramatic improvements following their introduction, looking at minicomputer speed, aircraft seat-miles-per-year capacity, and cement plant size. Others (Kurzweil 2005; Koh and Magee 2006; Koomey et al. 2011) have shown that new computers did not experience dramatic improvements in performance following their introduction, while I argue that large increases in seat-miles and plant size are merely an artifact of infrequent introductions of larger aircraft and larger manufacturing plants. Furthermore, neither of these measures of performance are relevant unless one discusses scaling, which Tushman and Anderson (1986) do not. Koh and Magee (2008) explicitly deny the existence of punctuated equilibrium in their analysis of energy storage technologies.

23. Lipsey, Carlaw, and Bekar (2005). Geometrical scaling is also different from network effects (Arthur 1994; Shapiro and Varian 1999) and increasing returns to R&D (Klepper 1996; Romer 1986).

24. Haldi and Whitcomb (1967), Levin (1977), Freeman and Louçã (2002), and Winter (2008). Rosenberg (1994, 198) estimates the increases in capital costs, with each doubling to be 60 percent.

25. See, for example, Sahal (1985), Lipsey, Carlaw, and Bekar (2005), and Winter (2008).

26. The first instance extends Lipsey, Carlaw, and Bekar's (2005) notion that the "ability to exploit [geometrical scaling] is dependent on the existing state of technology."

27. For example, the *Economist* devoted at least five articles to Christensen and his ideas in 2010 and 2011. However, other scholars suggest that Christensen's analysis may have exaggerated the challenges of disruptive innovations for incumbents (McKendrick, Haggard, and Doner 2000; King and Tucci 2002).

28. Even Christensen's newest book (Dyer, Gregersen, and Christensen 2011) implies these things by focusing solely on the skills needed for creating a low-end innovation and ignoring the improvements in performance and price that are needed for the innovation to displace the dominant technology and thus become a disruptive innovation.

29. See Bijker, Hughes, and Pinch (1997) for a discussion of the social construction of technologies, Schmookler (1966) for an analysis of R&D and demand, and others for analyses of the interaction between market needs and product designs (Clark 1985; Vincenti 1994; Arthur 2009).

30. One example of a change in user needs can be found in Tripsas (2008). New or existing users might also be the source of innovations (von Hippel 1988).

31. Lambert (2011).

32. Mowery and Rosenberg (1998) and Rosenberg (1982, 1994).

33. Nightingale (2008), Teece (2008), and others make similar arguments. For example, Freeman (1994) concludes that "the majority of innovations characterized as 'demand led' were actually relatively minor innovations along established trajectories"; Walsh (1984) and Fleck (1988) claim that supply-side factors drove innovation during its early stage in synthetic material, drugs, dyestuffs, and robotics.

34. Arrow (1962), Huber (1991), and March (1991).

35. An excellent treatment of this issue can be found in Gold (1981).

36. Although Arthur (2007) is one of the few to consider this time lag, others have considered the time lag between advances in science and the commercialization of the technology based on it. For example, see Klevorick et al. (1995), Kline and Rosenberg (1986), and Mansfield (1991).

37. Kaplan and Tripsas (2008) argue, and Dyer, Gregersen, and Christensen (2011) suggest, that cognitive bias is the main reason for any delay, while others largely ignore the issue (Anderson and Tushman 1990; Utterback 1994; Christensen and Bower 1996; Christensen 1997; Klepper 1997; Kaplan and Tripsas 2008). Exceptions include Levinthal (1998), who uses the notion of speciation to describe the "emergence" (Adner and Levinthal 2002) of new technologies, and Windrum (2005), who focuses on heterogeneity.

38. Although some (Gleick 2011) argue that a relay-based machine could have been constructed in the 19th century, this does not invalidate my logic that components were the main reason for the time lag.

39. Nathan Rosenberg and his colleagues (Rosenberg 1963, 1969; Kline and Rosenberg 1986; Mowery and Rosenberg 1998) emphasize the need for complementary technologies, while others emphasize a novel combination of technologies (Basalla 1988; Ayres 1988; Iansiti 1995; Fleming 2001; Hargadon 2003).

40. I am making a distinction between an ability to analyze and an ability to predict or forecast.

41. One exception is personal computers, where Microsoft's bundling of software is seen by some as anticompetitive. This is briefly mentioned in Chapters 7 and 8.

42. By related sectors, I refer to other types of magnetic storage and electronic systems.

43. Analyses of automobiles (Abernathy and Clark 1985), machine tools, electricity-generating equipment (Hughes 1983), and aircraft (Constant 1980; Vincenti 1994; Arthur 2009) suggest that novel combinations of components probably played a larger role in discontinuities than did improvements in single component types in the mechanical sector. On the other hand, there have been fewer discontinuities in the mechanical sector than in the electronics sector.

44. Modular design is a necessary but insufficient means for a component to have a large impact on the performance and costs of a system. For more on modular design, see Langlois (1992, 2003, 2007), Ulrich (1995), Sanchez and Mahoney (1996), and Baldwin and Clark (2000).

45. Malerba et al. (1999) make a similar argument but focus on radical innovations in components, while this book explains the emergence of discontinuities in terms of incremental innovations.

46. The notion of trade-offs is a fundamental property of indifference curves (Green and Wind 1973), Christensen's theory of disruptive innovation (Adner 2002, 2004; Adner and Zemsky 2005), and innovation frontiers (de Figueiredo and Kyle 2006).

47. Green and Wind (1973) and Adner (2002).

48. Windrum (2005) explicitly uses heterogeneity to examine discontinuities, while others (Levinthal 1998; Adner and Levinthal 2002) imply that heterogeneity is important.

49. Although Nordhaus (2009) makes the strongest argument about the problems of using the learning curve, others (Agarwal, Audretsch, and Sarkar 2007; Yang, Phelps, and Steensma 2010) have also noted these problems.

50. Afuah and Bahram (1995), Anderson and Tushman (1990), Tushman and Anderson (1986), and Utterback (1994).

51. Christensen (1997). For an opposing view, see King and Tucci (2002) and McKendrick, Haggard, and Doner (2000).

52. See, for example, Gort and Klepper (1982); Klepper and Graddy (1990); Agarwal and Gort (1996); Klepper (1997); Klepper and Simons (1997); and Tegarden, Hatfield, and Echols (1999).

53. Klepper (1996, 1997).

54. Klepper (1997) and Klepper and Thompson (2006). For vertical disintegration, also see Jacobides (2005), Jacobides and Winter (2005), Cacciatori and Jacobides (2005), and Jacobides and Billinger (2006).

55. This analysis builds from the research of Rohlfs (1974, 2001) and Tushman and Rosenkopf (1992).

56. Although Levitt and Dubner, in *SuperFreakonomics* (2009), also apply the concept of scaling to some solutions for global warming, they ignore the role of scaling in wind turbines, solar cells, and batteries.

57. Koh and Magee (2008) and personal communication with Chris Magee, May 13, 2011.

58. Kaku (2011), Stevenson (2011), and Diamandis and Kotler (2012).

59. One way to characterize a business model is in terms of value proposition, customer selection, method of value capture, scope of activities, and method of strategic control.

60. Drew (2011).

61. Slides for several chapters and those from presentations by students in a course based on this book are available at http://www.slideshare.net/Funk98/presentations.

PART I

1. http://www.cns-snc.ca/media/media/toocheap/toocheap.html.

2. Kahn and Wiener (1967) and Albright (2002).

CHAPTER 2

1. In order, these elements are similar to Dosi's emphasis on a "specific body of understanding," a "definition of the relevant problems to be addressed and the patterns of enquiry in order to address them," a "specific body of practice," and "the operative constraints on prevailing best practices and the problem-solving heuristics deemed promising for pushing back those constraints" (Dosi and Nelson 2010). The final item is consistent with combining components in novel ways, Hughes' (1983) characterization of reverse salients, and Arthur's (2009) characterization of how these reverse salients may lead to increasing complexity in an overall system.

2. See, for example, Arthur (2007).

3. Other methods of improvement such as automation, reducing the number of assembled parts, and modular design are probably equally applicable to all technologies and thus cannot explain why some technologies have experienced more improvements in cost and performance than have others. Improving yields and reducing energy consumption may be more important in "process" industries than in "assembly" industries, while combining parts may be more applicable in assembly industries. See Utterback (1994) and Chapter 3 for more details.

4. The historical aspects of this section largely draw on Cardwell (1995), Crump (2002), and Freeman and Soete (1997).

5. See, for example, http://en.wikipedia.org/wiki/James_Watt.

6. See, for example, Smil (2005, fig. 2.14) and Hirsh (1989).

7. See, for example, Smil (2007, 2010).

8. See, for example, http://en.wikipedia.org/wiki/Lift_%28force%29.

9. Smil (2005, fig. 3.14).

10. Intergovernmental Panel on Climate Change (2001).

11. The historical aspects of this section largely draw on Hughes (1983), Cardwell (1995), Freeman and Soete (1997), and Crump (2002).

12. See, for example, Smil (2010, fig. 4.1) or http://en.wikipedia.org/wiki/Energy_density.

13. Koh and Magee (2008).

14. Tarascon (2009).

15. The historical aspects of this section largely draw from Cardwell (1995), Numagami (1996, 1999), Castellano (2005), Crump (2002), and Orton (2009).

16. See, for example, Wikipedia on Geissler and Crooke's tubes. http://en.wikipedia.org/wiki/Geissler_tube and http://en.wikipedia.org/wiki/Crooke%27s_tube, respectively.

17. Nordhaus (1996).

18. An excellent history of these developments can be found in Orton (2009).

19. See, for example, Williamson (2010).

20. http://en.wikipedia.org/wiki/Phosphor.

21. See, for example, Numagami (1999).

22. See, for example, Castellano (2005), Numagami (1999), and Gay (2008).

23. Personal communication with Prof. Chris Magee (MIT), October 4, 2011.

24. The historical aspects of this section draw from a large number of sources, including Tilton (1971); Braun and Macdonald (1982); Lewis (1991); and Daniel, Mee, and Clark (1999).

25. Koh and Magee (2006). Similar data can be found in Koomey et al. (2011).

26. Data on the cost breakdown for many electronic products can be found in analyses done by iSuppli. For analyses of tablet computers, MP3 players, game consoles, televisions, e-book readers, and smart phones, see the following: www.isuppli.com/Teardowns/News/Pages/iSuppli-Does-Droid-Teardown-Finds-$18775-Bill-of-Materials-and-Manufacturing-Cost.aspx, www.isuppli.com/Teardowns/News/Pages/HP-TouchPad-Carries-$318-Bill-of-Materials.aspx, www.isuppli.com/Teardowns/News/Pages/Samsung-Galaxy-Tab-Carries-$205-Bill-of-Materials-iSuppli-Teardown-Reveals.aspx, www.informationweek.com/news/security/reviews/227700198, www.computerworld.com/s/article/9220471/Kindle_Fire_virtual_teardown_puts_cost_at209.63_about_10_above_retail, www.emsnow.com/newsarchives/archivedetails.cfm?ID=21201, www.isuppli.com/teardowns/pages/products.aspx, and www.emsnow.com/newsarchives/archivedetails.cfm?ID=15102.

27. Hasegawa and Takeya (2009).

28. The historical aspects of this section draw from sources such as Lewis (1991), Brock (1994), Coe (1996), and Garrard (1998).

29. For example, see the slides by Agrawal (2010).

30. See, for example, ISSCC (2010).

31. Amaya and Magee (2008) and Koh and Magee (2008).

32. A number of references address the role of supporting components along with a broader set of factors such as firm strategy and organization and infrastructure. See, for example, Freeman and Soete (1997) and Freeman and Louçã (2002).

33. See slides from my students and me for more data on several technologies. These slides (and those for several chapters here) are from a course based on this book and are available on SlideShare (http://www.slideshare.net/Funk98/edit_my_uploads), and discussed in my blog: http://jeffreyleefunk.blogspot.com/.

34. See Henderson's (1995) analysis of photolithographic equipment and how the semiconductor industry underestimated the potential for this technology. Also see Deutsch (2011) for a philosophical look at limits. I am indebted to Chris Magee for his insights on these issues.

CHAPTER 3

1. Chandler (1994) and Pratten (1971) emphasize that economies of scale exist more in continuous-flow than in other types of manufacturing and thus there are greater benefits from organizational scale in the former than in the latter.

2. See, for example, Stobaugh (1988); Morris, Campbell, and Roberts (1991); Utterback (1994); Freeman and Soete (1997); and Smil (2005, 2007).

3. For example, see Haldi and Whitcomb (1967); Axelrod, Daze, and Wickham (1968); Rosenberg (1994); Mannan (2005); Levin (1977); Freeman and Soete (1997); and Winter (2008).

4. If n is 0.65, the capital costs for the 5-million-ton plant are 0.0002 of those for the 0.1-million-ton plant on a per-unit basis (Axelrod, Daze, and Wickham 1968; Mannan 2005).

5. Enos (1962).

6. The height of a blast furnace wad doubled and its diameter increased by 30 percent between 1830 and 1913 (Smil 2005, fig. 4.1).

7. Lipsey, Carlaw, and Bekar (2005). For example, energy costs for ammonia production fell by one-half and by 70 percent for the best plants between 1920 and 2000 (Smil 2008, fig. 10.6).

8. Gold (1974).

9. See, for example, Smil (2005, fig. 4.12). Also see Electrochemistry Encyclopedia (http://electrochem.cwru.edu/encycl/art-a01-al-prod.htm) and aluminum statistics from the U.S. Geological Survey (minerals.usgs.gov/ds/2005/140/aluminum.pdf).

10. Rosenberg (1963, 1969) and Hounshell (1984).

11. It is difficult to increase machining and part-handling speeds particularly since individual parts must be handled and processed. Some of these and related issues are addressed by Rosenberg (1969).

12. Others have shown that product innovations do not give way to process innovations (Klepper and Simons 1997) and that economies of scale in R&D (Klepper 1997) better explain shakeouts than do dominant designs. ICs also highlight the problems with a distinction between product and process design because cost reductions are driven by reductions in the scale of product features that are enabled by changes in process design.

13. For example, see Utterback (1994) and Christensen (1997).

14. Chandler (1994).

15. Ayres (1988) and Lipsey, Carlaw, and Bekar (2005).

16. See, for example, Hounshell (1984). As an aside, the notion that improvements in manufacturing equipment depend on the use of this equipment to make parts for this equipment is also evident in computers; improvements in computers are needed in order in for semiconductor-manufacturing equipment to produce the computers' ICs.

17. Cardwell (1995), Crump (2002), McClellan and Dorn (2006), and Dosi and Nelson (2010).

18. Ayres (1988), Cardwell (1995), Crump (2002), and McClellan and Dorn (2006).

19. David (1990).

20. Hirsh (1989) gives 1973; others (Freeman and Soete 1997; Lipsey, Carlaw, and Bekar 2005; Munson 2005) cite different years.

21. Cardwell (1995), Crump (2002), and McClellan and Dorn (2006).

22. Lipsey, Carlaw, and Bekar (2005).

23. von Tunzelmann (1978) and www.western-marine.com/pdf/Honda_Marine_Retail_Price_list_2010.pdf.

24. Such a calculation assumes that production volumes are the same for all sizes of engines and that there are no limits to increasing the engines' scale. Since small engines probably have much higher production volumes than larger engines, the price data likely underestimates the benefits from scaling while the extrapolations likely overestimate them.

25. See Hirsh (1989) and in particular figures 16 through 21. Increasing the voltage in transmission systems dramatically reduced the energy losses in long-distance transmission; without low energy losses, it would have been difficult to benefit from the geometrical scaling in generating stations (Hirsh 1989; Munson 2005; Smil 2010).

26. Data on capital cost per output is from Hirsh (1989). Edison's Pearl Street Station Plant in 1880 was about 100 kilowatts. Benefits from geometrical scaling can also be seen in the price per kilowatt of existing diesel generators. For example, the price of a large Cummins engine (2250 kW) is less than 20 percent that of a small one (e.g., 7 kW) on a per-unit output basis. See the data on http://www.generatorjoe.net/store.asp.

27. Most of these combined-cycle turbines were made possible by advances in jet engines (Munson 2005).

28. Intergovernmental Panel on Climate Change (2001, chap. 7).

29. See, for example, Brian Arthur's (2009) analysis of jet engines, where dramatic increases in complexity were needed as large increases in scale were implemented.

30. Miller and Sawyers (1968); Lipsey, Carlaw, and Bekar (2005); and Winter (2008).

31. As with engines, there is greater demand for smaller aircraft and ships than smaller ones and thus differences in demand do not drive these differences in price per output. See UN Conference on Trade and Development (UNCTAD 2006) and http://www.boeing.com/commercial/cmo/. The smaller benefits from scaling in aircraft than from scaling in oil tankers may be because of the increasing cost of composites and other engineered materials in the largest aircraft. This was also found to be a problem in wind turbines (see Chapter 8).

32. Glaeser and Kohlhase (2004).

33. Freeman and Soete (1997), Cardwell (1995), Crump (2002), and McClellan and Dorn (2006).

34. Kurzweil (2005).

35. Patterson (2011); Christensen, Johnson, and Horn (2008); and Pisano (2006).

36. Feynman (1959).

37. Moore (1963), Tilton (1971), Braun and Macdonald (1982), Riordan and Hoddeson (1997), and Flamm (2004).

38. Studies have found that yields are lower for the ICs near the edge than at the center of the wafer in spite of the fact that the equipment is made substantially wider than the wafer and this part of the equipment is essentially wasted.

39. ICKnowledge (2009).

40. The newest fabrication facilities cost more than $3 billion, and it is estimated that only six producers can afford these facilities (*Financial Times* 2009).

41. See, for example, Foster (1986). Data on R&D expenditures can be found in Leckie (2005).

42. See, for example, Henderson (1995), Kurzweil (2005), Lim (2009), and *Financial Times* (2009).

43. Increases in disk size or tape width may also lead to slightly lower cost-per-bit storage. For example, the price of a 500-megabyte drive in early 2011 was $79.99 for 3.5 inch and $99.99 for an equivalent 2.5 inch. Furthermore, doubling the capacity of a 3.5 inch to 1 terabyte increased the price by only about 25 percent, to $99.99. See http://www.seagate.com/www/en-us/products/internal-storage/.

44. Smith (1988), Malerba et al. (1999), Kurzweil (2005), and Kressel and Lento (2007).

45. For example, see Grosch (1953, 1975), Knight (1968), and Ein-Dor (1985).

46. Ein-Dor (1985) argues that larger computers had higher markups, but subsequent analyses have concluded that these higher markups were due to higher development, sales, and distribution costs (Langlois 1992; Christensen 1997; Malerba et al. 1999).

47. For example, see the Constructive Cost Model for software, which estimates the cost of developing software as a function of lines of software code to a power between 1.05 and 1.20. See http://en.wikipedia.org/wiki/Cocomo.

48. Carr (2008).

49. See, for example, den Boer, (2005). Other evidence for similarities between these types of equipment is that leading providers of semiconductor and LCD (and solar) manufacturing equipment are the same firms.

50. Gay (2008).

51. Like IC wafers, large LCD substrates have smaller edge effects than do small substrates since LCD production equipment must be much wider than the substrate for consistent processing across the substrate. Thus, that extra width as a percentage of substrate width declines as substrate width increases.

52. Keshner and Arya (2004).

53. http://www.docstoc.com/docs/53390734/Flat-Panel-Display-Market-Outlook.

54. Keshner and Arya (2004).

55. *Economist* (2012).

56. For example, see Arrow (1962), Argote and Epple (1990), Huber (1991), March (1991), Ayres (1992), and Utterback (1994).

57. Klepper (1997) also shows how such a change from product to process innovations did not occur in automobiles and televisions.

58. This can also be analyzed in terms of integral versus modular design. Benefits from increasing the scale of a complex product require integral design, while benefits from increasing the scale of production equipment require modular design.

59. And by the theory that a discontinuity initially leads to *dramatic* improvements in performance and price (Tushman and Anderson 1986).

60. Christensen (1997).

61. Christensen, Craig, and Hart (2001); Christensen, Johnson, and Horn (2008); and Christensen, Grossman, and Hwang (2008).

62. Dyer, Gregersen, and Christensen (2011).

PART II

1. Henderson and Clark (1990), which draws on Abernathy and Clark (1985).

2. Abernathy and Clark (1985).

3. Christensen (1997).

4. Tushman and Murmann (1998) and Murmann and Frenken (2006).

5. de Figueiredo and Kyle (2006).

6. The drivers of performance for many systems were analyzed in Chapter 2 and are further analyzed in Chapters 4, 5, and 6. In Chapters 4, 5, and 6, it is shown that

single types of components had a stronger impact on system performance and cost than did novel combinations of components. Defining novel component combinations is inherently difficult. At one extreme, any change in a design can be interpreted as a novel combination and thus can define every new product. But do these changes drive improvements in performance and cost, or do changes in a single component drive them? Part II considers systems for which improvements in system performance depended more on improvements in component performance than on novel combinations of components.

7. Rosenberg (1963, 1969), Basalla (1988), Ayres (1988), Iansiti (1995), Mowery and Rosenberg (1998), and Fleming (2001).

8. Moore (1965).

9. Green and Wind (1979).

10. Windrum (2005) directly focuses on heterogeneity, while others indirectly address it using, for example, the notion of speciation (Levinthal 1998) to describe the "emergence" (Adner and Levinthal 2002) of new technologies.

11. Adner (2002, 2004) and Adner and Zemsky (2005).

12. de Figueiredo and Kyle (2006).

CHAPTER 4

1. This section draws primarily on Brock (1975), Flamm (1988), Ceruzzi (1998), Smith (1988), and van den Ende and Dolfsma (2005).

2. Smith (1988).

3. This section draws primarily on Brock (1975), Flamm (1988), Ceruzzi (1998), Smith (1988), and van den Ende and Dolfsma (2005).

4. This section draws primarily on Smith (1988), Flamm (1988), Inglis (1991), Pugh (1995), Pugh and Aspray (1996), Steinmueller (1996), Ceruzzi (1998), van den Ende and Dolfsma (2005), and Kurzweil (2005, 67).

5. Pugh and Aspray (1996).

6. Kurzweil (2005, 67).

7. Flamm (1988), Pugh and Aspray (1996), and Ceruzzi (1998).

8. This section draws primarily on Rifkin and Harrar (1988), Malerba (1985), Flamm (1988), Smith (1988), Christensen (1997), Ceruzzi (1998), and Baldwin and Clark (2000).

9. Mowery (1996).

10. One reason that mainframe computers were not used for these applications was that IBM wanted to maintain compatibility among all of its computers following introduction of the System/360 series in 1964.

11. This section draws primarily on Steffens (1994), Jackson (1997), Christensen (1997), Adner and Levinthal (2001), and Campbell-Kelly (2003).

12. Please note that I am not arguing that performance was more important than price or that the importance of performance initially rose rather than dropped over time, as others have shown (Adner and Levinthal 2001).

13. See, for example, Adner (2002).

14. This section draws primarily on Ceruzzi (1998), Rau and Fisher (1993), and Khazam and Mowery (1994).

15. This section draws primarily on Steffens (1994), Butter and Pogue (2002), Funk (2004), and Crossan and Mark (2005).

16. Steffens (1994).

17. Initially firms such as Apple and Microsoft focused on applications for which ICs could not meet a minimum threshold of performance. For example, it took six minutes to input a single address, and recognition of hand-written words did not exceed 80 percent in Apple's Newton, which was released in 1993. Microsoft's PDA had similar problems.

18. The fact that PCs had lower costs per processing speed because they used high-volume standard ICs suggests that firms should have expected a vertically disintegrated structure to eventually emerge for the industry. Vertical disintegration is not addressed until Chapter 7.

CHAPTER 5

1. This chapter is largely based on Funk (2009a).
2. This section largely draws on Daniel, Mee, and Clark (1999).
3. See, for example, Read and Welch (1976) and Clark (1999a).
4. This section largely draws on Daniel, Mee, and Clark (1999).
5. Trade-offs between other variables are addressed in Funk (2009a).
6. Millard (1995).
7. Millard (1995), Engel (1999), and Gooch (1999).
8. Millard (1995) and Clark (1999b).
9. Clark (1999b).
10. Sanderson and Uzumeri (1995).
11. This section primarily draws on Rosenbloom and Freeze (1985); Rosenbloom and Cusumano (1987); Inglis (1991); Cusumano, Mylonadis, and Rosenbloom (1992); Mallinson (1999); and Sadashige (1999).
12. Mallinson (1999) and Inglis (1991).
13. Rosenbloom and Cusumano (1987) and Cusumano, Mylonadis, and Rosenbloom (1992).
14. This section primarily draws on Grindley (1995), Sugaya (1999), and Johnstone (1999).
15. This section primarily draws on Sugaya (1999, fig. 12.7).
16. This section primarily draws on Millard (1995), Grindley (1995), Watkinson (1999), and Esener et al. (1999).
17. This section primarily draws on Inglis (1991), Sadashige (1999), and Esener et al. (1999).
18. Sadashige (1999).
19. Esener et al. (1999).
20. http://ixbtlabs.com/articles/storagetapes/index.html.
21. http://www.physorg.com/news183390187.html.

CHAPTER 6

1. Early versions of this chapter can be found in Funk (2008a, 2008b).

2. Henderson (1988) and Brown and Linden (2009).

3. Moore (1965) and Tuomi (2002).

4. Much of this section summarizes information from Tilton (1971), Braun and Macdonald (1982), Malerba (1985), Swann (1986), Turley (2003), and Flamm (2004).

5. Those who praise the innovativeness of microprocessor are primarily focusing on the unexpected applications that have emerged for them outside of computers.

6. See, for example, Deutsch (2011).

7. This section draws primarily on Moore (1963), Tilton (1971), Braun and Macdonald (1982), Riordan and Hoddeson (1997), Waits (2000), and Bassett (2002).

8. This section draws primarily on Tilton (1971), Reid (1985), and Riordan and Hoddeson (1997).

9. Moore (1965).

10. This section draws primarily on Majumder (1982), Ernst and O'Connor (1992), Malerba (1985), Swann (1986), and Bassett (2002).

11. Although not shown in this chapter, Peter Swann (1986) uses data on power dissipation (in his figure 4.2) and gate delay (in his figure 4.3) to construct the evolution of an innovation frontier for transistors (in his figure 4.4) in terms of gate delay and gate power dissipation between 1960 and 1980, and thus partly to show how this transition occurred.

12. This section draws primarily on Malerba (1985), Borrus (1987), Jackson (1997), and Aspray (1997).

13. As discussed in Chapter 4, the diffusion of laptop computers depended on improvements in a variety of components such as displays and ICs. See Yinug (2007).

14. This section draws primarily on Borrus (1987), Walker (1992), Linden and Somaya (2003), Thomke (2003), Turley (2003), and Rowen (2004).

15. With standard cell designs, which are the most "customized" of the ASICs, design engineers select predesigned blocks (which have increased in complexity as the number of transistors on a chip has increased) and then create special-purpose "masks" for their manufacture. The term *SoC* (system on a chip) is now used when these predesigned blocks include microprocessors and large blocks of memory. With so-called gate arrays, designers customize a standard array of transistors by choosing the connections between the transistors and thus the final mask layer of an IC. With programmable logic devices (PLDs) or field programmable gate arrays (FPGAs), design engineers can customize either of these "standard" products in a matter of minutes by connecting specific "fuses" (Borrus 1987; Linden and Somaya 2003; Thomke 2003; Rowen 2004). They can do this because high-level functions such as high-speed logic, memory, and processors are embedded into each chip. The difference between PLDs and FPGAs is that FGPAs can be reprogrammed even after they have been assembled into a product and installed in the field (Turley 2003).

16. For example, see Rowen (2004, fig. 8.2).

17. Some of the foundational concepts and technologies for programmable logic arrays, gates, and logic blocks are founded in patents awarded to David W. Page and

 LuVerne R. Peterson in 1985 (http://en.wikipedia.org/wiki/Field-programmable_gate_array).

18. This section draws primarily on Walker (1992) and Turley (2003).

19. Langlois (1992) and Takahashi (1999).

20. Henderson (1995).

21. Lim (2009).

22. Kurzweil (2005), Teo (2010), and Humphries (2010).

23. There is a huge body of literature on these topics and on nanotechnology. The most useful sources I have found include Clark Ngyuen's and Alex Zettl's lectures at Berkeley (http://webcast.berkeley.edu/) and Baldwin (2010); Kaku (2011); Teo (2010); and Khademhosseini, Vacanti, and Langer. (2009). Their ideas are also summarized in the slides I use to help my students analyze the future of these technologies, which are available at http://www.slideshare.net/Funk98/presentations.

24. Malerba (1985); Borrus (1987); Rappaport and Halevi (1991); Stoelhorst (1997); and Macher, Mowery, and Hodges (1999).

25. Rosenberg (2004).

CHAPTER 7

1. Afuah and Bahram (1995), Anderson and Tushman (1990), Tushman and Anderson (1986), Leonard-Barton (1992), and Utterback (1994).

2. Christensen (1997). For an opposing viewpoint, see King and Tucci (2002) and McKendrick, Haggard, and Doner (2000).

3. See, for example, Gort and Klepper (1982); Klepper and Graddy (1990); Agarwal and Gort (1996); Klepper (1997); Klepper and Simons (1997); and Tegarden, Hatfield, and Echols (1999). The number of automobile manufacturers peaked at about 300, while the number of television and PC manufacturers peaked at about 50 each.

4. See, for example, Gruber (1994); Cusumano, Mylonadis, and Rosenbloom (1992); Flamm (1996); and Funk (2002).

5. See, for example, Utterback (1994), Suarez and Utterback (1995), Shane (2004), and Baron and Shane (2005).

6. Klepper (1997), Klepper and Simons (1997), and Klepper and Thompson (2006).

7. http://www.census.gov/epcd/susb/latest/us/US54.HTM and *Economist* (1993).

8. Braun and Macdonald (1982), Borrus (1987), Saxenian (1994), Angel (1994), and Dow Jones (2006).

9. For example, Chandler (1994) emphasizes vertical integration in his analyses of the late 19th and early 20th centuries.

10. See, for example, Langlois (1992, 2003, 2007); Baldwin and Clark (2000); Arora, Fosuri, and Gambardella (2001); Steinmueller (2003); Jacobides (2005); Jacobides and Winter (2005); Cacciatori and Jacobides (2005); and Jacobides and Billinger (2006).

11. Jacobides (2005), Jacobides and Winter (2005), Cacciatori and Jacobides (2005), and Jacobides and Billinger (2006).

12. Langlois (2003, 2007); Arora, Fosuri, and Gambardella (2001); Caves (2002); and Steinmueller (2003).

13. Jacobides (2005) and Jacobides and Winter (2005).

14. See, for example, Ulrich (1995), Sanchez and Mahoney (1996), Brusoni and Prencipe (2001), and Baldwin and Clark (2000).

15. Farrell and Saloner (1985), Katz and Shapiro (1985, 1994), Arthur (1994), and Shapiro and Varian (1999).

16. Flamm (1988), Rifkin and Harrar (1988), Anderson and Tushman (1990), and Ceruzzi (1998).

17. Campbell-Kelly (2003).

18. Steinmueller (1996) and Cortada (2005).

19. Steinmueller (1996).

20. *Datamation* magazine presented data each June on the top 50 (and for some years the top 100) IT firms in terms of revenues for the years 1975 through 1996.

21. See Funk (2012b).

22. Ceruzzi (1998), Campbell-Kelly (2003), and Cusumano (2004).

23. Ceruzzi (1998) and Moore (1999).

24. Like the other data presented previously, this percentage is based on data from the June issues of *Datamation* for the years 1975 through 1996.

25. This analysis is based on data from *Datamation* magazine.

26. for data on the numbers of firms, see, among others, Braun and Macdonald (1982), Borrus (1987), Saxenian (1994), Angel (1994), and Dow Jones (2006). For data on shakeouts in specific segments, see, for example, Gruber (1994) and Flamm (1996).

27. Tilton (1971), Braun and Macdonald (1982), and Steinmueller (1987).

28. ASICs, semi-custom ICs designed for specific applications, include standard cell, gate array, and programmable logic array (Thomke 2003). Although some firms have designed and manufactured ASICs for special applications, such as those for the military, since the 1960s, sales did not begin to grow until firms emerged that only supplied design tools and databases and left the detailed design to the systems manufacturers. ASSPs are standard IC chips designed for a specific system/product and often a specific module in that system/product.

29. Walker (1992); ICE (1996); Baldwin and Clark (2000); and Macher, Mowery, and Simcoe (2002).

30. The number U.S. foundries is actually quite small. Most are based in Taiwan or Singapore. Also, because of the large economies of scale required, only a small number of them can co-exist.

31. Brown and Linden (2009) contrast the economies of scale for foundries, design houses, and other firms.

32. See, for example, Funk (2009b, 2012a) and Christensen, Verlinden, and Westerman (2002).

33. Langlois (2003, 2007); Arora, Fosuri, and Gambardella (2001); Caves (2002); and Steinmueller (2003).

CHAPTER 8

1. This analysis builds from the research of Rohlfs (1974, 2001) and Tushman and Rosenkopf (1992).

2. This typology draws on Funk (2010).

3. This is consistent with Tushman and Rosenkopf (1992): "the greater the number of subsystems and the greater the number of interface and interdependence demands, the greater the numbers of dimensions of merit that must be adjudicated. The more complex the system, the more political, social, and community dynamics operate to resolve trade-offs between alternative technical choices." Also see Bijker, Hughes, and Pinch (1997); Hobday (1998); and Geels (2002, 2004).

4. Rohlfs (1974, 2001), Katz and Shapiro (1994), and Shapiro and Varian (1999).

5. Cusumano, Mylonadis, and Rosenbloom (1992) and Levy (1989).

6. Rogers (2003) and Agarwal and Bayus (2002).

7. Vernon (1971) and Beise (2004).

8. Vernon (1971), Hughes (1983), and Chandler (1994).

9. See, for example, Rosenbloom and Freeze (1985), Sanderson and Uzumeri (1995), and Windrum (2005).

10. von Burg (2001), Peterson (1995), Rohlfs (2001), and Hart (2004).

11. Rohlfs (2001).

12. Rohlfs (1974, 2001).

13. Millard (1995).

14. Langlois and Robertson (1992).

15. Shapiro and Varian (1999); Nelson, Peterhansl, and Sampat (2004); Grindley (1995); and Rohlfs (2001).

16. Langlois (1992), Grindley (1995), Rohlfs (2001), and Bresnahan (2007).

17. Tilton (1971), Hughes (1983), Flamm (1988), Cowan (1990), Tushman and Murmann (1998), Mowery and Rosenberg (1998), and Kirsch (2000).

18. Kenney (1986) and Pisano (2006).

19. Patterson (2011) and Christensen, Johnson, and Horn (2008).

20. Tilton (1971), Hughes (1983), Flamm (1988), Pugh (1995), Tushman and Murmann (1998), Mowery and Rosenberg (1998), Kirsch (2000), and Rohlfs (2001).

21. Flamm (1988) and Shapiro and Varian (1999).

22. Briggs and Burke (2002), Noam (1991), Funk (2002), Kogut (2003), and Campbell-Kelly (2003).

23. Briggs and Burke (2002), Noam (1991), Funk (2002, 2007a), Kogut (2003), and Fransman (2002).

24. Garrard (1998) and Funk (2002).

25. Mowery and Simcoe (2002) and Kogut (2003).

26. Kogut (2003).

27. Fransman (2002) and Funk (2007a).

PART IV

1. Kaku (2011), Diamandis and Kotler (2012), and Stevenson (2011).

CHAPTER 9

1. This section draws primarily on various papers and books, including Numagami (1996, 1999), den Boer (2005), and Castellano (2005), and on analyses done by my students on 3D displays (see http://www.slideshare.net/Funk98/presentations).

2. Ng and Funk (2011).

3. This section draws primarily on Garrard (1998); Gilder (2002); Levinthal (1998); Rabaey (2011); and Yoffie, Yin, and Kind (2004).

4. This analysis draws primarily on a presentation by my 2011 graduate students on cognitive radio (see http://www.slideshare.net/Funk98/presentations).

5. This section draws primarily on MacKie-Mason and Varian (1994), Abbate (1999), Segaller (1998), Huurdeman (2003), and Kressel and Lento (2007).

6. MacKie-Mason and Varian (1994).

7. This is largely based on a presentation by Rajan Walia at the National University of Singapore in March 2011 and on conversations with Aaron Danner of the National University of Singapore.

8. This section draws primarily on Kenney (2003), Okin (2005), and Tapscott and Williams (2006).

9. This section draws primarily on Carr (2008) and on a project by one of my graduate students, Maynard Kang.

10. Carr (2008).

11. Lohr (2011).

12. See, for example, "The Cost of Implementing ERP," http://www.networksasia.net/content/costs-implementing-erp.

13 http://www.ft.com/intl/cms/s/0/19f14406-a118-11e0-9a07-00144feabdc0.html#axzz1RxXTZ6ps.

14. See, for example, the analysis by Carnegie Mellon's Software Engineering Institute. http://www.sei.cmu.edu/sos/research/cloudcomputing/cloudbarriers.cfm.

15. This section draws primarily on Funk (2004, 2006, 2007a, 2007b, 2012a).

16. See, for example, Funk (2006, 2007a, 2007b, 2012a).

17. This section draws primarily on a presentation by my 2011 graduate students on human-computer interfaces (see http://www.slideshare.net/Funk98/presentations).

18. The Future of Human-Computer Interfaces, http://www.slideshare.net/Funk98/presentations.

19. Suzuki (2010).

CHAPTER 10

1. The U.S. government expects to spend $150 billion through 2019 on clean energy, of which less than $5 billion is expected to involve research on and development of solar cells and wind turbines. This observation was made by Dan Arvizu in a presentation at the National University of Singapore, in November 2010, entitled "Moving toward a Clean Energy Future." Also see Ball (2010). Germany has spent $130 billion on subsidies and this enables it to now receive 0.3 percent of its energy from the sun (Lomborg 2012).

2. This section largely draws on Narasimalu and Funk (2011).

3. Although the best-fit regression curve has an R^2 of 95percent, the raw data suggests that most of this variability comes from rotor diameters greater than 70 meters.

4. A variety of health problems also exist for those living close to wind turbines (Pierpoint 2009).

5. Zervos (2008).

6. http://www.norway.org.au/News_and_events/Business/The-worlds-largest-wind-turbine-to-be-built-in-Norway/.

7. I am indebted to Lee Branstetter for his question about the relative sizes of the Singapore Flyer and wind turbines. For more detail on wind turbine size, see Zervos (2008) and Nemet (2009).

8. http://www.newenergyworldnetwork.com/renewable-energy-news/by-technology/wind/uk-innovator-wind-power-launches-10mw-aerogenerator-x-wind-turbine-design.html.

9. www.economist.com/science/tq/displaystory.cfm?story_id=9249242.

10. Nemet's (2006) analysis of wind turbine implementations concluded that public funding has not encouraged design innovations.

11. In addition to conversations with many people in the solar industry, including those at Nanosolar, Applied Materials, and REC, and my reading of many consulting reports and books (Perlin 1999; Bradford 2006; Nelson 2006), my thoughts in this section primarily rely on research by two of my former graduate students, Mary Joyce Sagayno and Shukai Chen.

12. Interestingly, many of these countries are in North Africa and the Middle East, where much of the world's oil is found. Some argue that Europe's electricity may end up being produced in North Africa and delivered via transmission lines that are laid under the Mediterranean Sea.

13. Shah et al. (1999) and Nemet (2006).

14. A more precise characterization of band gaps would include a discussion of voltage and current. Photons striking materials with a high band gap produce higher voltages but lower current than do photons striking materials with a low band gap. For the distribution of the solar spectrum, see http://org.ntnu.no/solarcells/pics/chap2/Solar_Spectrum.png.

15. See Nelson (2006, chap. 2).

16. Reliability is a major issue with most solar users, and few are optimistic about these systems. On the other hand, given the falling cost of ICs, it is likely that the cost of the control portion of these systems will fall over time. But how fast and by how much? What is the maximum threshold of price before they will be cost-effective in solar panels, and how long will it take before their price drops below it?

17. http://en.wikipedia.org/wiki/Solar_cells.

18. Other sources provide cost data consistent with that shown in Figure 10.1. For example, see Evans (2009) and Hamilton (2010).

19. Brown (2010).

20. Some argue that the toxicity of cadmium and a shortage of telluride are bigger reasons for their low production. This may be the case, but, from a policy standpoint, if these are problems, I would argue that governments should probably subsidize R&D for environmental controls for cadmium and subsidize a search for telluride rather than merely emphasize demand-based subsidies and thus the production of silicon solar cells.

21. http://www.greentechmedia.com/articles/read/inside-optisolars-grand-ambitions-6029/.

22. This section primarily relies on a presentation by and conversations with Palani Balaya, including a recommended paper (Tarascon 2010).

23. Tarascon (2009).

24. Personal communication with Chris Magee, May 19, 2011. For flywheels, see http://www.economist.com/node/21540386.

25. Lowe et al. (2010).

26. *Economist* (2011a).

27. While conventional batteries separate the two electrodes with a thick barrier, nanoscale batteries place the electrodes close to each other with nanowires and other nanodevices. This can increase the rate of exchange between the electrodes and thus the energy and power densities. The reason is that as the diameter of the electrode or catalyst particles is reduced, the ratio of surface area to volume goes up, and thus the rate of exchange between particles increases. See http://www.sciencedaily.com/releases/2010/09/100914151043.htm, ScientificComputing.com, and

http://www.scientificcomputing.com/news-DS-Building-Better-Batteries-from-the-Nanoscale-Up-121010.aspx.

28. See http://www.amazon.com/Sun-Bottle-Strange-History-Thinking/dp/B001JT6FHS. Also see Kaku (2011).

29. See note 1.

30. See, for example, Nemet (2006, 2009) and Jacobson and Delucchi (2009).

31. See Pontin (2010).

CHAPTER 11

1. For learning curves, see Arrow (1962), Ayres (1992), Huber (1991), Argote and Epple (1990), and March (1991). For improvements in processes, see Abernathy and Utterback (1978), Utterback (1994), Christensen (1997), and Adner and Levinthal (2001).

2. Moore (1965).

3. Nelson and Winter (1982) and Christensen (1997).

4. Moore (1963) and Waits (2000).

5. Other research has found similar results. For example, Daniel Levinthal has shown how technological developments in industries such as electric power and fixed-line telecommunication led to wireless technology and its diffusion across many applications. Mary Tripsas has shown how "radical technology, from the perspective of the typesett[ing] industry, had been developed incrementally in other industries for many years." In both cases, we can interpret the sources of discontinuities as incremental improvements in components: in wireless telecommunication, improvements in components in transmitters and amplifiers; in the typesetting industry, improvements in the components in optics, film, relays, motors, CRT tubes, lasers, and software. An important conclusion here is that existing industries often drive the improvements in components that make new industries possible, and thus assessing the potential of new industries requires us to consider the extent to which technologies, equipment, and processes can be borrowed from other industries. See Levinthal (1998) and Tripsas (2008, table 3).

6. Also see Nordhaus (2009).

7. For example, one of the world's leading academic conferences on innovation, DRUID (Danish Research Unit on Industrial Dynamics), debated this issue at its 2008 meeting. The following resolution was discussed and narrowly rejected: "Let it be

resolved that this conference believes that in order to improve the quality and impact of empirical research on industry dynamics we should discourage the use of patent data as a measure of innovation." See http://www.druid.dk/index.php?id=20.

8. As Christensen (1997) states, "Whether the technology was radical or incremental, expensive or cheap, software or hardware, component or architecture, competence-enhancing or competence-destroying, the pattern was the same. When faced with sustaining technology change that gave existing customers something more and better in what they wanted, the leading practitioners of the prior technology led the industry in the development and adoption of the new." On the other hand, his most recent book (Dyer, Gregersen, and Christensen 2011) refers to sustaining innovations as incremental innovations and thus contributes to the confusion.

9. McKendrick, Haggard, and Doner (2000) and King and Tucci (2002).

10. Such an interpretation is also supported by an emphasis by many papers on disruptive innovation on finding new niches and not on the circumstances under which a new technology might replace an existing one. See the *Journal of Product Innovation Management* (Special Issue) 25 (4), July 2008.

11. For example, a heavy emphasis on disruptive innovation in previous courses caused more than one-third of the students in one of my courses to write the following: "Technology A is inferior to the existing technology and thus is a disruptive innovation and will replace the existing technology." They seemed to believe that no further analysis was required.

12 Yu and Hang (2011).

13. Bresnahan and Trajtenberg (1995).

14. See, for example, Cowen (2011) and *Economist* (2011b).

15. Drew (2011).

16. Deutsch (2011) and Kaku (2011).

17. Slides for several chapters and those from presentations by my students in a course based on this book are available at http://www.slideshare.net/Funk98/presentations.

References

Abbate, J. 1999. *Inventing the Internet.* Cambridge, MA: MIT Press.

Abernathy, W., and K. Clark. 1985. "Innovation: Mapping the Winds of Creative Destruction." *Research Policy* 14 (1): 3–22.

Abernathy, W., and J. Utterback. 1978. "Patterns of Innovation in Technology." *Technology Review* 80 (7): 40–47.

Adner, R. 2002. "When Are Technologies Disruptive? A Demand-Based View of the Emergence of Competition." *Strategic Management Journal* 23 (8): 667–688.

———. 2004. "A Demand-Based Perspective on Technology Lifecycles." *Advances in Strategic Management* 21: 25–43.

Adner, R., and D. Levinthal. 2001. "Demand Heterogeneity and Technology Evolution: Implications for Product and Process Innovation." *Management Science* 47 (5): 611–628.

———. 2002. "The Emergence of Emerging Technologies." *California Management Review* 45 (1): 50–66.

Adner, R., and P. Zemsky. 2005. "Disruptive Technologies and the Emergence of Competition." *RAND Journal of Economics* 36 (2): 229–254.

Afuah, A., and N. Bahram. 1995. "The Hypercube of Innovation." *Research Policy* 24 (1): 51–76.

Agarwal, R., D. Audretsch, and M. Sarkar. 2007. "The Process of Creative Construction: Knowledge Spillovers, Entrepreneurship, and Economic Growth." *Strategic Entrepreneurship Journal* 1 (3–4): 263–286.

Agarwal, R., and B. Bayus. 2002. "The Market Evolution and Sales Takeoff of Product Innovations." *Management Science* 48 (8): 1024–1041.

Agarwal, R., and M. Gort. 1996. "The Evolution of Markets and Entry, Exit, and Survival of Firms." *Review of Economics and Statistics* 78 (3): 489–498.

Agrawal, G. 2010. "Fiber-Optic Communication Systems." Accessed August 10, 2011. www.master-photonics.org/uploads/media/Govind_Agrawal1.pdf.

Albright, R. 2002. "What Can Past Technology Forecasts Tell Us about the Future?" *Technological Forecasting and Social Change* 69 (5): 443–464.

Alexander, C. 1964. *Notes on the Synthesis of Form*. Cambridge, MA: Harvard University Press.

Amaya, M., and C. Magee. 2008. "The Progress in Wireless Data Transport and Its Role in the Evolving Internet." Technical report, Massachusetts Institute of Technology, Cambridge, MA. Accessed September 10, 2011. web.mit.edu/~cmagee/www/.../24-theprogressinwirelessdatatransportc.pdf.

Anderson, P., and M. Tushman. 1990. "Technological Discontinuities and Dominant Designs: A Cyclical Model of Technological Change." *Administrative Science Quarterly* 35 (4): 604–633.

Angel, D. 1994. *Restructuring for Innovation: The Remaking of the U.S. Semiconductor Industry*. New York: Guilford.

Argote, L., and D. Epple. 1990. "Learning Curves in Manufacturing." *Science* 247 (4945): 920–924.

Arora, A., A. Fosuri, and A. Gambardella. 2001. *Markets and Technology: The Economics of Innovation and Corporate Strategy*. Cambridge, MA: MIT Press.

Arrow, K. 1962. "The Economic Implications of Learning by Doing." *Review of Economic Studies* 29 (3): 155–173.

Arthur, B. 1994. *Increasing Returns and Path Dependence in the Economy*. Ann Arbor: University of Michigan Press.

———. 2007. "The Structure of Invention." *Research Policy* 36 (2): 274–287.

———. 2009. *The Nature of Technology*. New York: Free Press.

Aspray, W. 1997. "The Intel 4004 Microprocessor: What Constituted Innovation?" *IEEE Annals of the History of Computing* 19 (3): 4–15.

Axelrod, L., R. Daze, and H. Wickham. 1968. "The Large Plant Concept." *Chemical Engineering Progress* 64 (7): 17.

Ayres, R. 1988. "Barriers and Breakthroughs: An 'Expanding Frontiers' Model of the Technology-Industry Life Cycle." *Technovation* 7 (2): 87–115.

———. 1992. "Experience and the Life Cycle: Some Analytic Implications." *Technovation* 12 (7): 465–486.

Azevedo, I., et al. 2009. "The Transition to Solid-State Lighting." *Proceedings of the IEEE* 97: 481–510.

Balconi, M., S. Brusoni, and L. Orsenigo. 2010. "In Defence of the Linear Model: An Essay." *Research Policy* 39 (1): 1–13.

Baldwin, C., and K. Clark. 2000. *Design Rules*. Cambridge, MA: MIT Press.

Baldwin, D. 2010. "New and Emerging Technologies in Electronics." Accessed November 23, 2011. www.smta.org/chapters/files/Uppermidwest_VendorExpo_New_and_Emerging_Technologies_051811.pdf.

Ball, J. 2010. "Clean Energy Sources: Sun, Wind and Subsidies." *Wall Street Journal*, January 10. Accessed November 3, 2010. http://online.wsj.com/article/SB126290539750320495.html.

Baron, R., and S. Shane. 2005. *Entrepreneurship: A Process Perspective*. Mason, OH: South-Western.

Basalla, G. 1988. *The Evolution of Technology*. Cambridge: Cambridge University Press.

Bassett, R. 2002. *To the Digital Age*. Baltimore: Johns Hopkins University Press.

Beise, M. 2004. "Lead Markets: Country-Specific Drivers of the International Diffusion of Innovations." *Research Policy* 33 (6–7): 997–1018.

Bijker, W., T. Hughes, and T. Pinch. 1997. *The Social Construction of Technological Systems: New Directions in the Sociology and History of Technology*. 6th ed. Cambridge, MA: MIT Press.

Borrus, M. 1987. *Competing for Control: America's Stake in Microelectronics*. New York: Ballinger.

Bradford, T. 2006. *Solar Revolution*. Cambridge, MA: MIT Press.

Braun, E., and S. Macdonald. 1982. *Revolution in Miniature: The History and Impact of Semiconductor Electronics*. Cambridge: Cambridge University Press.

Bresnahan, T. 2007. "Creative Destruction in the PC Industry." In *Perspectives on Innovation*, edited by F. Malerba and S. Brusoni, 105–140. Cambridge: Cambridge University Press.

Bresnahan, T., and M. Trajtenberg. 1995. "General Purpose Technologies 'Engines of Growth'?" *Journal of Econometrics* 65 (1): 83–108.

Briggs, A., and P. Burke. 2002. *A Social History of the Media: From Gutenberg to the Internet*. Cambridge: Polity.

Brock, G. 1975. *The U.S. Computer Industry: A Study of Market Power*. Cambridge, MA: Ballinger.

———. 1981. *The Telecommunications Industry: The Dynamics of Market Structure*. Cambridge, MA: Harvard University Press.

———. 1994. *Telecommunication Policy for the Information Age: From Monopoly to Competition*. Cambridge, MA: Harvard University Press.

Brown, C., and G. Linden. 2009. *Chips and Change: How Crisis Reshapes the Semiconductor Industry*. Cambridge, MA: MIT Press.

Brown, L. 2010. "Solar Cell Production Increases 51%." TreeHugger.com. Accessed March 15, 2011. http://www.treehugger.com/files/2010/09/solar-cell-production-increases-51-percent.php#.

Brusoni, S., and A. Prencipe. 2001. "Unpacking the Black Box of Modularity: Technologies, Products and Organizations." *Industrial and Corporate Change* 10 (1): 179–205.

Bussey, G. 1990. *Wireless, the Crucial Decade: History of the British Wireless Industry, 1924–1934*. London: Peter Peregrinus.

Butter, A., and D. Pogue. 2002. *Piloting Palm: The Inside Story of Palm, Handspring, and the Birth of the Billion-Dollar Handheld Industry*. New York: Wiley.

Cacciatori, E., and M. Jacobides. 2005. "The Dynamic Limits of Specialization: Vertical Integration Reconsidered." *Organization Studies* 26 (12): 1851–1883.

Campbell-Kelly, M. 2003. *From Airline Reservations to Sonic the Hedgehog: A History of the Software Industry*. Cambridge, MA: MIT Press.

Cardwell, D. 1995. *Wheels, Clocks, and Rockets: A History of Technology*. New York: Norton.

Carr, N. 2008. *The Big Switch: Rewiring the World, from Edison to Google*. New York: Norton.

Castellano, J. 2005. *Liquid Gold: The Story of Liquid Crystal Displays and the Creation of an Industry*. Singapore: World Scientific.

Caves, R. 2002. *The Creative Industries*. Cambridge, MA: Harvard University Press.

Ceruzzi, P. 1998. *A History of Modern Computing*. Cambridge, MA: MIT Press.

Chandler, A. 1994. *Scale and Scope: The Dynamics of Industrial Capitalism*. Boston: Belknap.

Christensen, C. 1997. *The Innovator's Dilemma*. Boston: Harvard Business School Press.

Christensen, C., and J. Bower. 1996. "Customer Power, Strategic Investment, and the Failure of Leading Firms." *Strategic Management Journal* 17 (3): 197–218.

Christensen, C., T. Craig, and S. Hart. 2001. "The Great Disruption." *Foreign Affairs* 80 (2): 80–95.

Christensen, C., J. Grossman, and J. Hwang. 2008. *The Innovator's Prescription*. New York: McGraw-Hill.

Christensen, C., C. Johnson, and M. Horn. 2008. *Disrupting Class*. New York: McGraw-Hill.

Christensen, C., M. Verlinden, and G. Westerman. 2002. "Disruption, Disintegration, and the Dissipation of Differentiability." *Industrial and Corporate Change* 11 (5): 955–993.

Clark, K. 1985. "The Interaction of Design Hierarchies and Market Concepts in Technological Evolution." *Research Policy* 14 (5): 235–251.

Clark, M. 1999a. "Steel Tape and Wire Recorders." In *Magnetic Recording: The First 100 Years*, edited by E. Daniel, C. Mee, and M. Clark, 30–46. New York: IEEE Press.

———. 1999b. "Product Diversification." In *Magnetic Recording: The First 100 Years*, edited by E. Daniel, C. Mee, and M. Clark, 92–109. New York: IEEE Press.

Coe, L. 1996. *Wireless Radio: A History*. Jefferson, NC: McFarland.

Constant, E. 1980. *The Origins of Turbojet Revolution*. Baltimore: Johns Hopkins University Press.

Cortada, J. 2005. *The Digital Hand*. Vol. 2. *How Computers Changed the Work of American Financial, Telecommunications, Media and Entertainment Industries*. New York: Oxford University Press.

Cowan, R. 1990. "Nuclear Power Reactors: A Study in Technological Lock-In." *Journal of Economic History* 50: 541–567.

Cowen, T. 2011. *The Great Stagnation*. New York: Penguin.

Crossan, M., and K. Mark. 2005. "Apple Inc.: iPods and iTunes." Harvard Business School Case #905M46.

Crump, T. 2002. *Science: As Seen through the Development of Scientific Instruments*. London: Constable and Robinson.

Cusumano, M. 2004. *The Business of Software*. New York: Free Press.

Cusumano, M., Y. Mylonadis, and R. Rosenbloom. 1992. "Strategic Maneuvering and Mass-Market Dynamics: The Triumph of VHS over Beta." *Business History Review* 66 (1): 51–94.

Daniel, E., C. Mee, and M. Clark. 1999. *Magnetic Recording: The First 100 Years*. New York: IEEE Press.

David, P. 1990. "The Dynamo and the Computer: An Historical Perspective on the Modern Productivity Paradox." *American Economic Review* 80 (2): 355–361.

de Figueiredo, J., and M. Kyle. 2006. "Surviving the Gales of Creative Destruction: The Determinants of Product Turnover." *Strategic Management Journal* 27 (3): 241–264.

den Boer, W. 2005. *Active Matrix Liquid Crystal Displays*. New York: Elsevier.

Deutsch, D. 2011. *The Beginning of Infinity*. New York: Penguin.

Diamandis, P., and S. Kotler. 2012. *Abundance: The Future Is Better Than You Think*. New York: Free Press.

Dosi, G. 1982. "A Suggested Interpretation of the Determinants and Directions of Technical Change." *Research Policy* 11 (3): 147–162.

Dosi, G., and R. Nelson. 2010. "Technical Change and Industrial Dynamics as Evolutionary Processes." In *Handbook of the Economics of Innovation*, edited by B. Hall and N. Rosenberg, 51–126. Burlington, MA: Academic Press.

Dow Jones. 2006. *Venture Capital Industry Report*. New York: Dow Jones.

Drew, C. 2011. "Why Science Majors Change Their Minds." *New York Times*, November 4.

Dyer, J., H. Gregersen, and C. Christensen. 2011. *The Innovator's DNA: Mastering the Five Skills of Disruptive Innovators*. Boston: Harvard Business School Press.

Economist. 1993. "Harsh New World." February 27, pp. 7–11.

———. 2011a. "The Power of the Press." January 20, p. 73.

———. 2011b. "Review of *The Great Stagnation*, by Tyler Cowen." January 27. http://www.economist.com/blogs/freeexchange/2011/01/growth_2.

2012. "Television Making: Cracking Up." January 21, p. 66.

Ein-Dor, P. 1985. "Grosch's Law Revisited." *Communications of the ACM* 28 (2): 142–151.

Eisenhardt, K. 1989. "Building Theories from Case Study Research." *Academy of Management Review* 14 (4): 532–550.

Engel, F. 1999. "The Introduction of the Magnetophon." In *Magnetic Recording: The First 100 Years*, edited by E. Daniel, C. Mee, and M. Clark, 47–71. New York: IEEE Press.

Enos, J. 1962. *Petroleum Progress and Profits*. Cambridge, MA: MIT Press.

Ernst, D., and D. O'Connor. 1992. *Competing in the Electronic Industry*. Paris: Organisation for Economic Co-operation and Development.

Esener, S., M. Kryder, W. Doyle, M. Keshner, M. Mansuripur, and D. Thompson. 1999. "The Future of Data Storage Technologies." Panel Report, World Technology Evaluation Center, Baltimore. http://www.wtec.org/loyola/hdmem/toc.htm.

Evans, P. 2009. "Solar Panel Industry Achieves Holy Grail—$1 per Watt Grid-Parity." *gizmag*, March 3. Accessed July 16, 2010. http://www.gizmag.com/solar-panel-1-per-watt-grid-parity/11143/.

Farrell, J., and G. Saloner. 1985. "Standardization, Compatibility, and Innovation." *RAND Journal of Economics* 16 (1): 70–83.

Feynman, R. 1959. "There's Plenty of Room at the Bottom: An Invitation to Enter a New Field of Physics." Talk given at the annual meeting of the American Physical Society. Accessed December 3, 2009. http://www.zyvex.com/nanotech/feynman.html.

Financial Times. 2009. "Moore's Law Hits Economic Limits." July 21.

Fischer, C. 1987. "The Revolution in Rural Telephony, 1900–1920." *Journal of Social History* 21 (1): 5–26.

Flamm, K. 1988. *Creating the Computer: Government, Industry, and High Technology*. Washington, DC: Brookings Institution.

———. 1996. *Mismanaged Trade?* Washington DC: Brookings Institution.

———. 2004. "Economic Growth and Semiconductor Productivity." In *Productivity and Cyclicality in Semiconductors: Trends, Implications, and Questions—Report of a Symposium*, edited by D. Jorgenson and C. Wessner, 43–59. Washington, DC: National Research Council.

Fleck, J. 1988. "Innofusion or Diffusation? The Nature of Technological Developments in Robotics." Working paper, Edinburgh PICT Centre, Edinburgh University.

Fleming, L. 2001. "Recombinant Uncertainty in Technological Search." *Management Science* 47 (1): 117–132.

Foster, R. 1986. *The Attacker's Advantage*. New York: Basic Books.

Fransman, M. 2002. *Telecoms in the Internet Age: From Boom to Bust to . . . ?* Oxford: Oxford University Press.

Freeman, C. 1994. "The Economics of Technical Change." *Cambridge Journal of Economics* 18 (5): 463–514.

Freeman, C., and F. Louçã. 2002. *As Time Goes By: From the Industrial Revolutions to the Information Revolution*. Oxford: Oxford University Press.

Freeman, C., and L. Soete. 1997. *The Economics of Industrial Revolution*. Cambridge, MA: MIT Press.

Funk, J. 2002. *Global Competition between and within Standards: The Case of Mobile Phones*. London: Palgrave.

———. 2004. *Mobile Disruption: The Technologies and Applications Driving the Mobile Internet*. New York: Wiley.

———. 2006. "The Future of Mobile Phone–Based Intranet Applications: A View from Japan." *Technovation* 26 (12): 1337–1346.

———. 2007a. "Solving the Startup Problem in Western Mobile Internet Markets." *Telecommunications Policy* 31 (1): 14–30.

———. 2007b. "Mobile Shopping: Visions from Japan." *Technological Forecasting and Social Change* 61 (3): 341–356.

———. 2008a. "Components, Systems, and Technological Discontinuities: Lessons from the IT Sector." *Long Range Planning* 41 (5): 555–573.

———. 2008b. "Systems, Components and Technological Discontinuities: The Case of the Semiconductor Industry." *Industry and Innovation* 15 (4): 411–433.

———. 2009a. "Systems, Components, and Technological Discontinuities: The Case of Magnetic Recording and Playback Equipment." *Research Policy* 38 (7): 1079–1216.

———. 2009b. "The Emerging Value Network in the Mobile Phone Industry: The Case of Japan and Its Implications for the Rest of the World." *Telecommunications Policy* 33 (1): 4–18.

———. 2010. "Complexity, Network Effects, and Industry Formation: A Survey of Industries." *Industry and Innovation* 17 (5): 483–502.

———. 2012a. "Multiple Standards and Critical Masses, and the Formation of New Industries: The Case of the Japanese Mobile Internet." *European Journal of Innovation Management* 15 (1): 4–26.

———. 2012b. "The Unrecognized Connection between Vertical Disintegration and Entrepreneurial Opportunities." *Long Range Planning* 45 (1): 41–59.

Garrard, G. 1998. *Cellular Communications: Global Market Development*. Boston: Artech House.

Gay, C. 2008. "Applied Materials: Accelerating Solar Power Cost Reduction." Presentation given on June 12, San Jose, CA.

Geels, F. 2002. "Technological Transitions as Evolutionary Reconfiguration Processes: A Multi-Level Perspective and a Case-Study." *Research Policy* 31 (8–9): 1257–1274.

———. 2004. "From Sectoral Systems of Innovation to Socio-Technical Systems: Insights about Dynamics and Change from Sociology, and Institutional Theory." *Research Policy* 33 (6–7): 897–920.

Gilder, G. 2002. *Telecosm*. 2nd ed. New York: Touchstone.

Glaeser, E., and J. Kohlhase. 2004. "Cities, Regions and the Decline of Transport Costs." *Papers in Regional Science* 83 (1): 197–228.

Gleick, J. 2011. *The Information: A History, a Theory, a Flood*. New York: Pantheon.

Gold, B. 1974. "Evaluating Scale Economies: The Case of Japanese Blast Furnaces." *Journal of Industrial Economics* 23 (1): 1–18.

———. 1981. "Changing Perspectives on Size, Scale, and Returns: An Interpretive Survey." *Journal of Economic Literature* 19 (1): 5–33.

Gooch, B. 1999. "Building on the Magnetophon." In *Magnetic Recording: The First 100 Years*, edited by E. Daniel, C. Mee, and M. Clark, 72–91. New York: IEEE Press.

Gort, M., and S. Klepper. 1982. "Time Paths in the Diffusion of Product Innovations." *Economic Journal* 92 (367): 630–653.

Gould, S., and N. Eldredge. 1977. "Punctuated Equilibria: The Tempo and Mode of Evolution Reconsidered." *Paleobiology* 3 (2): 115–151.

Green, P., and Y. Wind. 1973. *Multi-Attribute Decisions in Marketing: A Measurement Approach*. Hinsdale, IL: Dryden.

Grindley, P. 1995. *Standards Strategy and Policy: Cases and Stories*. Oxford: Oxford University Press.

Grosch, H. 1953. "High Speed Arithmetic: The Digital Computer as a Research Tool." *Journal of the Optical Society of America* 43 (April): 4.

———. 1975. "Grosch's Law Revisited." *Computerworld* 8 (16): 24.

Gruber, H. 1994. *Learning and Strategic Product Innovation: Theory and Evidence for the Semiconductor Industry*. New York: North Holland.

Haldi, J., and D. Whitcomb. 1967. "Economies of Scale in Industrial Plants." *Journal of Political Economy* 75 (4): 373–385.

Hamilton, T. 2010. "Solar Cell Maker Gets a $400-Million Boost." *Technology Review*, July 14. Accessed July 16, 2010. http://www.technologyreview.com/energy/25801/.

Hargadon, A. 2003. *How Breakthroughs Happen: The Surprising Truth about How Companies Innovate*. Boston: Harvard Business School Press.

Hart, J. 2004. *Technology, Television, and Competition: The Politics of Digital TV*. Cambridge: Cambridge University Press.

Hasegawa, T., and J. Takeya. 2009. "Organic Field-Effect Transistors Using Single Crystals." *Science and Technology of Advanced Materials* 10: 1–16.

Henderson, R. 1988. "The Failure of Established Firms in the Face of Technical Change: A Study of Photolithographic Alignment Equipment." Unpublished PhD dissertation, Harvard University, Cambridge, MA.

———. 1995. "Of Life Cycles Real and Imaginary: The Unexpectedly Long Old Age of Optical Lithography." *Research Policy* 24 (4): 631–643.

Henderson, R., and K. Clark. 1990. "Architectural Innovation: The Reconfiguration of Existing Product Technologies and the Failure of Established Firms." *Administrative Science Quarterly* 35 (1): 9–30.

Hirsh, R. 1989. *Technology and Transformation in the Electric Utility Industry*. Cambridge: Cambridge University Press.

Hobday, M. 1998. "Product Complexity, Innovation and Industrial Organization." *Research Policy* 26 (6): 689–710.

Hounshell, D. 1984. *From the American System to Mass Production, 1800–1932: The Development of Manufacturing Technology in the United States*. Baltimore: Johns Hopkins University Press.

Huber, G. 1991. "Organizational Learning: The Contributing Processes and the Literatures." *Organization Science* 2 (1): 71–87.

Hughes, T. 1983. *Networks of Power*. Baltimore: Johns Hopkins University Press.

Humphries, C. 2010. "A Moore's Law for Genetics." *Technology Review*, March–April. http://www.technologyreview.com/biomedicine/24590/.

Huurdeman, A. 2003. *The Worldwide History of Telecommunications*. New York: Wiley.

Iansiti, M. 1995. "Technology Integration: Managing Technological Evolution in a Complex Environment." *Research Policy* 24 (4): 521–542.

ICKnowledge. 2009. "Rising Costs of Fabrication." Accessed December 7, 2009. http://www.icknowledge.com/economics/fab_costs.html.

Inglis, A. 1991. *Behind the Tube: A History of Broadcasting Technology and Business*. Boston: Focal.

Integrated Circuit Engineering (ICE). 1996. *North American Company Profiles, Integrated Circuit Engineering, Market Report*. smithsonianchips.si.edu/ice/cd/PROF96/NORTHAM.PDF.

Intergovernmental Panel on Climate Change. 2001. *Aviation and the Global Atmosphere*. Cambridge: Cambridge University Press.

ISSCC. 2010. "Trends Report." 2010 International Solid-State Circuits Conference. Accessed April 17, 2012. isscc.org/doc/2010/ISSCC2010_TechTrends.pdf.

Jackson, T. 1997. *Inside Intel: Andy Grove and the Rise of the World's Most Powerful Chip Company*. New York: Dutton.

Jacobides, M. 2005. "Industry Change through Vertical Disintegration: How and Why Markets Emerged in Mortgage Banking." *Academy of Management Journal* 48 (3): 465–498.

Jacobides, M., and S. Billinger. 2006. "Designing the Boundaries of the Firm: From 'Make, Buy or Ally' to the Dynamic Benefits of Vertical Architecture." *Organization Science* 17 (2): 249–261.

Jacobides, M., and S. Winter. 2005. "The Co-evolution of Capability and Transaction Costs: Explaining the Institutional Structure of Production." *Strategic Management Journal* 26 (5): 395–413.

Jacobson, M., and M. Delucchi. 2009. "A Path to Sustainable Energy by 2030." *Scientific American*, November 1.
Johnstone, B. 1999. *We Were Burning: Japanese Entrepreneurs and the Forging of the Electronic Age*. New York: Basic Books.
Kahn, H., and A. Wiener. 1967. *The Year 2000: A Framework for Speculation on the Next Thirty-Three Years*. London: Macmillan.
Kaku, M. 2011. *Physics of the Future*. New York: Doubleday.
Kaplan, S., and M. Tripsas. 2008. "Thinking about Technology: Applying a Cognitive Lens to Technical Change." *Research Policy* 37 (5): 790–805.
Katz, M., and C. Shapiro. 1985. "Network Externalities, Competition, and Compatibility." *American Economic Review* 75 (3): 424–440.
———. 1994. "Systems Competition and Network Effects." *Journal of Economic Perspectives* 8 (2): 93–115.
Kenney, M. 1986. *Biotech: The University-Industry Complex*. New Haven, CT: Yale University Press.
———. 2003. "The Growth and Development of the Internet in the United States." In *The Global Internet Economy*, edited by B. Kogut, 69–108. Cambridge, MA: MIT Press.
Keshner, M., and R. Arya. 2004. "Study of Potential Cost Reductions Resulting from Super-Large Scale Manufacturing of PV Modules." Final Report NREL/SR-520-36846, National Renewable Energy Laboratory, Golden, CO.
Khademhosseini, A., J. Vacanti, and R. Langer. 2009. "How to Grow New Organs." *Scientific American*, May 4.
Khazam, J., and D. Mowery. 1994. "The Commercialization of RISC: Strategies for the Creation of Dominant Designs." *Research Policy* 23 (1): 89–102.
King, A., and C. Tucci. 2002. "Incumbent Entry into New Market Niches: The Role of Experience and Managerial Choice in the Creation of Dynamic Capabilities." *Management Science* 48 (2): 171–186.
Kirsch, D. 2000. *The Electric Vehicle and the Burden of History*. New Brunswick, NJ: Rutgers University Press.
Klepper, S. 1996. "Entry, Exit, Growth, and Innovation over the Product Life Cycle." *American Economic Review* 86 (3): 562–583.
———. 1997. "Industry Life Cycles." *Industrial and Corporate Change* 6 (1): 145–181.
———. 2007. "Disagreements, Spinoffs, and the Evolution of Detroit as the Capital of the U.S. Automobile Industry." *Management Science* 53 (4): 616–631.
———. 2010. "The Origin and Growth of Industry Clusters: The Making of Silicon Valley and Detroit." *Journal of Urban Economics* 67 (1): 15–32.
Klepper, S., and E. Graddy. 1990. "The Evolution of New Industries and the Determinants of Market Structure." *RAND Journal of Economics* 21 (1): 27–44.
Klepper, S., and K. Simons. 1997. "Technological Extinctions of Industrial Firms: An Inquiry into Their Nature and Causes." *Industrial and Corporate Change* 6 (2): 379–460.
Klepper, S., and P. Thompson. 2006. "Submarkets and the Evolution of Market Structure." *RAND Journal of Economics* 37 (4): 861–886.
Klevorick, A., R. Levin, R. Nelson, and S. Winter. 1995. "On the Sources and Significance of Inter-Industry Differences in Technological Opportunities." *Research Policy* 24 (2): 185–205.

Kline, S., and N. Rosenberg. 1986. "An Overview on Innovation." In *The Positive Sum Strategy*, edited by R. Landau and N. Rosenberg, 275–305. Washington, DC: National Academy Press.

Knight, K. 1968. "Evolving Computer Performance 1963–1967." *Datamation* 14 (1): 31–35.

Kogut, B. 2003. *The Global Internet Economy*. Cambridge, MA: MIT Press.

Koh, H., and C. Magee. 2006. "A Functional Approach for Studying Technological Progress: Application to Information Technologies." *Technological Forecasting and Social Change* 73 (9): 1061–1083.

———. 2008. "A Functional Approach for Studying Technological Progress: Extension to Energy Technology." *Technological Forecasting and Social Change* 75 (6): 735–758.

Koomey, J., S. Berard, M. Sanchez, and H. Wong. 2011. "Implications of Historical Trends in the Electrical Efficiency of Computing." *IEEE Annals of the History of Computing* 33 (3): 46–54.

Kressel, H., and T. Lento. 2007. *Competing for the Future: How Digital Innovations Are Changing the World*. New York: Cambridge University Press.

Kuhn, T. 1970. *The Structure of Scientific Revolutions*. Chicago: University of Chicago Press.

Kurzweil, R. 2005. *The Singularity Is Near*. New York: Penguin.

Lambert, R. 2011. "Its Camp Is Gone But the Occupy Movement Will Grow." *Financial Times*, November 15.

Langlois, R. 1992. "External Economics and Economic Progress: The Case of the Microcomputer Industry." *Business History* 66 (1): 1–50.

———. 2003. "The Vanishing Hand: The Changing Dynamics of Industrial Capitalism." *Industrial and Corporate Change* 12 (2): 351–385.

———. 2007. *The Dynamics of Industrial Capitalism: Schumpeter, Chandler, and the New Economy*. New York: Taylor and Francis.

Langlois, R., and P. Robertson. 1992. "Networks and Innovation in a Modular System: Lessons from the Microcomputer and Stereo Component Industries." *Research Policy* 21 (4): 297–313.

Leckie, R. 2005. "Semiconductor Equipment and Materials: Funding the Future of R&D." Accessed December 7, 2009. www.semi.org/cms/groups/public/documents/web_content/p036612.pdf.

Leonard-Barton, D. 1992. "Core Capabilities and Core Rigidities: A Paradox in Managing New Product Development." *Strategic Management Journal* 13 (Special Issue): 111–125.

Levin, R. 1977. "Technical Change and Optimal Scale: Some Implications." *Southern Economic Journal* 2: 208–221.

Levinthal, D. 1998. "The Slow Pace of Rapid Technological Change: Gradualism and Punctuation in Technological Change." *Industrial and Corporate Change* 7 (2): 217–247.

Levitt, S., and S. Dubner. 2009. *Super Freakonomics*. New York: Morrow.

Levy, M. 1989. *The VCR Age: Home Video and Mass Communication*. London: Sage.

Lewis, T. 1991. *The Empire of the Air: The Men Who Made Radio*. New York: HarperCollins.

Lim, K. 2009. "The Many Faces of Absorptive Capacity: Spillovers of Copper Interconnect Technology for Semiconductor Chips." *Industrial and Corporate Change* 18 (6): 1249–1284.

Linden, G., and D. Somaya. 2003. "System-on-a-Chip Integration in the Semiconductor Industry: Industry Structure and Firm Strategies." *Industrial and Corporate Change* 12 (3): 545–576.

Lipsey, R., K. Carlaw, and C. Bekar. 2005. *Economic Transformations*. New York: Oxford University Press.

Lohr, S. 2011. "Microsoft Takes to Cloud to Defend Its Office Business." *New York Times*, June 28.

Lomborg, B. 2012. "Germany's Sunshine Daydream." Accessed February 16, 2012. http://www.project-syndicate.org/commentary/germany-s-sunshine-daydream.

Lowe, M., S. Tokuoka, T. Trigg, and G. Gereffi. 2010. "Lithium-Ion Batteries for Electric Vehicles." Center on Globalization, Governance and Competitiveness, Duke University, Durham, NC.

Macher, J., D. Mowery, and D. Hodges. 1999. "Semiconductors." In *US Industry in 2000: Studies in Competitive Performance*, edited by D. Mowery, 245–286. Washington, DC: National Academy Press.

Macher, J., D. Mowery, and T. Simcoe. 2002. "e-Business and Disintegration of the Semiconductor Industry Value Chain." *Industry and Innovation* 9 (3): 155–181.

MacKie-Mason, J., and H. Varian. 1994. "Economic FAQs about the Internet." *Journal of Economic Perspectives* 8 (3): 75–96.

Majumder, B. 1982. *Innovations, Product Developments, and Technology Transfers: An Empirical Study of Dynamic Competitive Advantage: The Case of Electronic Calculators*. Washington, DC: University Press of America.

Malerba, F. 1985. *The Semiconductor Business: The Economics of Rapid Growth and Decline*. London: Pinter.

Malerba, F., R. Nelson, L. Orsenigo, and S. Winter. 1999. "History-Friendly Models of Industry Evolution: The Computer Industry." *Industrial and Corporate Change* 8 (1): 3–40.

Mallinson, J. 1999. "The Ampex Quadruplex Recorders." In *Magnetic Recording: The First 100 Years*, edited by E. Daniel, C. Mee, and M. Clark, 153–169. New York: IEEE Press.

Mannan, S. 2005. *Lee's Loss Prevention in the Process Industries*. Vol. 1. Burlington, MA: Elsevier Butterworth-Heinemann.

Mansfield, E. 1991. "Academic Research and Industrial Innovation." *Research Policy* 20 (1): 1–12.

March, J. 1991. "Exploration and Exploitation in Organizational Learning." *Organization Science* 2 (1): 71–87.

McCallum, J. 2008. Accessed April 24, 2012. http://www.jcmit.com/cpu-performance.htm.

McClellan, J., and H. Dorn. 2006. *Science and Technology in World History*. Baltimore: Johns Hopkins University Press.

McKendrick, D., S. Haggard, and R. Doner. 2000. *From Silicon Valley to Singapore: Location and Competitive Advantage in the Hard Disk Drive Industry*. Stanford, CA: Stanford University Press.

Millard, A. 1995. *America on Record: A History of Recorded Sound*. New York: Cambridge University Press.

Miller, R., and D. Sawyers. 1968. *The Technical Development of Modern Aviation*. London: Routledge.

Misra, A., P. Kumar, M. Kamalasanan, and S. Chandra. 2006. "White Organic LEDs and Their Recent Advancements." *Semiconductor Science and Technology* 21 (7): R35–R47.

Molstad, R., D. Langlois, and D. Johnson. 2002. "Linear Tape Servo Writing Enables Increased Track Density." Data Storage. Accessed September 27, 2008. http://www.imationltd.co.uk/products/pdfs/Network_Tape_article_servo.pdf.

Moore, G. 1963. "Integrated Circuits." In *Microelectronics*, edited by E. Keonjian, 262–359. New York: McGraw-Hill.

———. 1965. "Cramming More Components onto Integrated Circuits." *Electronics* 38 (8): 114–117.

———. 1999. *Crossing the Chasm: Marketing and Selling High-Tech Products to Mainstream Customers*. New York: HarperBusiness.

———. 2004. "No Exponential Is Forever . . . ." Accessed May 16, 2006. ftp://download.intel.com/research/silicon/Gordon_Moore_ISSCC_021003.pdf.

Morris, P., W. Campbell, and H. Roberts. 1991. *Milestones in 150 Years of the Chemical Industry*. Cambridge: Royal Society of Chemistry.

Mowery, D. 1996. *The International Computer Software Industry*. New York: Oxford University Press.

Mowery, D., and N. Rosenberg. 1998. *Paths of Innovation*. New York: Cambridge University Press.

Mowery, D., and T. Simcoe. 2002. "Is the Internet a US Invention? An Economic and Technological History of Computer Networking." *Research Policy* 31 (8–9): 1369–1387.

Munson, R. 2005. *From Edison to Enron: The Business of Power and What It Means for the Future of Electricity*. New York: Praeger.

Murmann, P., and K. Frenken. 2006. "Toward a Systematic Framework for Research on Dominant Designs, Technological Innovations, and Industrial Change." *Research Policy* 35 (7): 925–952.

Narasimalu, S., and J. Funk. 2011. "Geometrical Scaling and Components: The Case of Wind Turbines." Paper presented at the IEEE International Technology Management Conference, San Jose, CA, June 27–30.

Nelson, G. 2006. *The Physics of Solar Cells*. London: Imperial College Press.

Nelson, R., A. Peterhansl, and B. Sampat. 2004. "Why and How Innovations Get Adopted: A Tale of Four Models." *Industrial and Corporate Change* 13 (5): 679–699.

Nelson, R., and S. Winter. 1982. *An Evolutionary Theory of Economic Change*. Cambridge, MA: Belknap/Harvard University Press.

Nemet, G. 2006. "Beyond the Learning Curve: Factors Influencing Cost Reductions in Photovoltaics." *Energy Policy* 34 (17): 3218–3232.

———. 2009. "Demand-Pull, Technology-Push, and Government-Led Incentives for Non-Incremental Technical Change." *Research Policy* 38 (5): 700–709.

Ng, P., and J. Funk. 2011. "Technology Paradigms and the Timing of Technological Discontinuities: The Case of Three-Dimensional Television." Working paper.

Nightingale, P. 2008. "Meta-Paradigm Change and the Theory of the Firm." *Industrial and Corporate Change* 17 (3): 533–583.

Noam, E. 1991. *Television in Europe*. New York: Oxford University Press.

Nordhaus, W. 1996. "Do Real-Output and Real-Wage Measures Capture Reality? The History of Lighting Suggests Not." In *The Economics of New Goods*, edited by T. Bresnahan, 27–70. Cambridge, MA: National Bureau of Economic Research.

———. 2009. "The Perils of the Learning Model for Modeling Endogenous Technological Change." Discussion Paper 1685, Cowles Foundation for Research in Economics, Yale University, New Haven, CT.

Numagami, T. 1996. "Flexibility Trap: A Case Analysis of US and Japanese Technological Choice in the Digital Watch Industry." *Research Policy* 25 (1): 133–162.

———. 1999. *History of Technological Innovation in TFT-LCD Display* [in Japanese]. Tokyo: Hakuto.

Okin, J. 2005. *The Internet Revolution*. Winter Harbor, ME: Ironbound.

O'Neil, P. 2003. "CMOS Technology and Business Trends: Can the Semiconductor Industry Afford to Continue Advancing." Accessed March 7, 2005. http://ewh.ieee.org/r5/denver/sscs/Presentations/2002.12.ONeill.pdf.

Orton, J. 2009. *Semiconductors and the Information Revolution*. New York: Academic Press.

Patterson, S. 2011. *The Quants*. New York: Crown Business.

Perlin, J. 1999. *From Space to Earth: The Story of Solar Electricity*. Cambridge, MA: Harvard University Press.

Peterson, M. 1995. "The Emergence of a Mass Market for Fax Machines." *Technology in Society* 17 (4): 469–482.

Pierpoint, N. 2009. *Wind Turbine Syndrome: A Report on a Natural Experiment*. Santa Fe: K-Selected Books.

Pisano, G. 2006. *Science Business*. Boston: Harvard Business School Press.

Pontin, J. 2010. "Q&A: Bill Gates, the Cofounder of Microsoft, Talks Energy, Philanthropy and Management Style. *Technology Review*, August 24. Accessed August 26, 2010. http://www.technologyreview.com/energy/26112/page1/.

Pratten, C. 1971. *Economies of Scale in Manufacturing Industry*. Cambridge: Cambridge University Press.

Pugh, E. 1995. *Building IBM: Shaping an Industry and Its Technology*. Cambridge, MA: MIT Press.

Pugh, E., and W. Aspray. 1996. "A History of the Information Machine." *IEEE Annals of the History of Computing* 18 (2): 70–76.

Rabaey, J. 2011. "Silicon Architectures for Wireless Systems—Part 1." Accessed April 23, 2012. http://www.eecs.berkeley.edu/~jan.

Rappaport, A. S., and S. Halevi. 1991. "The Computerless Computer Company." *Harvard Business Review* 69 (4): 69-80.

Rau, B., and J. Fisher. 1993. "Instruction-Level Parallel Processing: History, Overview and Perspective." *Journal of Supercomputing* 7 (1–2): 9–50.

Read, O., and W. Welch. 1976. *From Tin Foil to Stereo: Evolution of the Phonograph*. Indianapolis: Howard Sams/Bobbs-Merrill.

Reid, T. 1985. *The Chip: How Two Americans Invented the Microchip and Launched a Revolution*. New York: Simon and Schuster.

Rifkin, G., and G. Harrar. 1988. *The Ultimate Entrepreneur: The Story of Ken Olsen and Digital Equipment Corporation*. New York: Contemporary Books.

Riordan, M., and S. Hoddeson. 1997. *Crystal Fire: The Invention and Birth of the Information Age*. New York: Norton.

Rogers, E. 2003. *Diffusion of Innovations*. 5th ed. New York: Free Press.

Rohlfs, J. 1974. "A Theory of Interdependent Demand for a Communications Service." *Bell Journal of Economics and Management Science* 5 (1): 16–37.

———. 2001. *Bandwagon Effects in High-Technology Industries*. Cambridge, MA: MIT Press.

Romer, P. 1986. "Increasing Returns and Long-Run Growth." *Journal of Political Economy* 94 (5): 1002–1037.

Rosenberg, N. 1963. "Technological Change in the Machine Tool Industry, 1840–1910." *Journal of Economic History* 23 (4): 414–443.

———. 1969. "The Direction of Technological Change: Inducement Mechanisms and Focusing Devices." *Economic Development and Cultural Change* 18 (1): 1–24.

———. 1982. *Inside the Black Box: Technology and Economics*. New York: Cambridge University Press.

———. 1994. *Exploring the Black Box*. Cambridge: Cambridge University Press.

———. 2004. "Science and Technology: Which Way Does the Causation Run?" Presentation on the occasion of the opening of a new "Center for Interdisciplinary Studies of Science and Technology," Stanford University, Stanford, CA, November 1.

Rosenbloom, R., and M. Cusumano. 1987. "Technological Pioneering and Competitive Advantage: The Birth of the VCR Industry." *California Management Review* 29 (4): 51.

Rosenbloom, R., and K. Freeze. 1985. "Ampex Corporation and Video Innovation." *Research on Technological Innovation, Management and Policy* 2: 113–185.

Rowen, C. 2004. *Engineering the Complex SoC*. Upper Saddle River, NJ: Prentice Hall.

Sadashige, K. 1999. "Digital Video Recording." In *Magnetic Recording: The First 100 Years*, edited by E. Daniel, C. Mee, and M. Clark, 201–219. New York: IEEE Press.

Sahal, D. 1985. "Technological Guideposts and Innovation Avenues." *Research Policy* 14 (2): 61–82.

Sanchez, R., and J. Mahoney. 1996. "Modularity, Flexibility, and Knowledge Management in Product and Organization Design." *Strategic Management Journal* 17 (Special Issue): 63–76.

Sanderson, S., and M. Uzumeri. 1995. "Managing Product Families: The Case of the Sony Walkman." *Research Policy* 24 (5): 761–782.

Saxenian, A. 1994. *Regional Advantage: Culture and Competition in Silicon Valley and Route 128*. Cambridge, MA: Harvard University Press.

Schmidt-Nielsen, K. 1984. *Scaling: Why Is Animal Size So Important?* New York: Cambridge University Press.

Schmookler, J. 1966. *Invention and Economic Growth*. Cambridge, MA: Harvard University Press.

Segaller, S. 1998. *Nerds: A Brief History of the Internet*. New York: TV Books.

Shah, A., P. Torres, R. Tscharner, N. Wyrsch, and H. Keppner. 1999. "Photovoltaic Technology: The Case for Thin-Film Solar Cells." *Science* 285 (5428): 692–698.

Shane, S. 2004. *Finding Fertile Ground: Identifying Extraordinary Opportunities for New Ventures*. Philadelphia: Wharton School Publishing.

Shapiro, C., and H. Varian. 1999. *Information Rules*. Boston: Harvard Business School Press.

Sheats, J., et al. 1996. "Organic Electroluminescent Devices." *Science* 273 (5277): 884–888.

Signet Solar. 2007. "Glass and Module Size for Thin Film Solar." EPIA Workshop, November 22.

Simon, H. 1962. "The Architecture of Complexity." *Proceedings of the American Philosophical Society* 106 (6): 467–482.

Smil, V. 2005. *Creating the Twentieth Century*. New York: Oxford University Press.

———. 2007. *Transforming the Twentieth Century: Technical Innovations and Their Consequences*. New York: Oxford University Press.

———. 2008. *Energy in Nature and Society*. Cambridge, MA: MIT Press.

———. 2010. *Energy Transitions*. New York: Praeger.

Smith, R. 1988. "A Historical Overview of Computer Architecture." *IEEE Annals of the History of Computing* 10 (4): 277–303.

Steffens, J. 1994. *Newgames: Creating Competition in the PC Revolution*. New York: Pergamon.

Steinmueller, W. 1987. "Microeconomics and Microelectronics: Economic Studies of Integrated Circuit Technology." Unpublished dissertation, Stanford University, Stanford, CA.

———. 1996. "The U.S. Software Industry: An Analysis and Interpretive History." In *The International Computer Software Industry*, edited by D. Mowery, 15–52. New York: Oxford University Press.

———. 2003. "The Role of Technical Standards in Coordinating the Division of Labour in Complex System Industries." In *The Business of Systems Integration*, edited by A. Prencipe, A. Davies, and M. Hobday, 133–151. New York: Oxford University Press.

Stevenson, M. 2011. *An Optimist's Tour of the Future*. New York: Profile Books.

Stobaugh, R. 1988. *Innovation and Competition: The Global Management of Petrochemical Products*. Boston: Harvard Business School Press.

Stoelhorst, J. 1997. "In Search of a Dynamic Theory of the Firm." Unpublished PhD dissertation, University of Twente, Enschede, Netherlands.

Suarez, F., and J. Utterback. 1995. "Dominant Designs and the Survival of Firms." *Strategic Management Journal* 16 (6): 415–430.

Sugaya, H. 1999. "Consumer Video Recorders." In *Magnetic Recording: The First 100 Years*, edited by E. Daniel, C. Mee, and M. Clark, 182–200. New York: IEEE Press.

Suzuki, T. 2010. "Challenges of Image-Sensor." International Solid States Circuit Conference.

Swann, P. 1986. *Quality Innovation: An Economic Analysis of Rapid Improvements in Microelectronic Components*. Westport, CT: Greenwood.

Takahashi, D. 1999. "3-D Daytona 500." *Electronics Business* (November): 118–126.

Tapscott, D., and A. Williams. 2006. *Wikinomics: How Mass Collaboration Changes Everything*. New York: Portfolio.

Tarascon, J. 2009. "Batteries for Transportation Now and in the Future." Paper presented at Energy 2050, Stockholm, October 19–20.

———. 2010. "Key Challenges in Future Li-Battery Research." *Philosophical Transactions of the Royal Society* 368 (1923): 3227–3241.

Teece, D. 2008. "Dosi's Technological Paradigms and Trajectories: Insights for Economics and Management." *Industrial and Corporate Change* 17 (3): 467–484.

Tegarden, L., D. Hatfield, and A. Echols. 1999. "Doomed from the Start: What Is the Value of Selecting a Future Dominant Design?" *Strategic Management Journal* 20 (6): 495–518.

Teo, K. 2010. "Point-of-Care Diagnostics and Drug Discovery Tools: An Approach from Bioelectronics Perspective." Accessed September 25, 2010. semi.org/cms/groups/public/documents/web.../ctr_029927.pdf.

Thomke, S. 2003. *Experimentation Matters: Unlocking the Potential of New Technologies for Innovation.* Boston: Harvard Business School Press.

Tilton, J. 1971. *The International Diffusion of Technology: The Case of Semiconductors.* Washington, DC: Brookings Institution.

Tripsas, M. 2008. "Customer Preference Discontinuities: A Trigger for Radical Technological Change." *Managerial Decision Economics* 29 (2–3): 79–97.

Tuomi, I. 2002. "The Lives and Death of Moore's Law." *First Monday* 7 (11). Accessed April 23, 2012. http://firstmonday.org/htbin/cgiwrap/bin/ojs/index.php/fm/rt/printerFriendly/1000/921.

Turley, J. 2003. *The Essential Guide to Semiconductors.* Upper Saddle River, NJ: Prentice Hall.

Tushman, M., and P. Anderson. 1986. "Technological Discontinuities and Organizational Environment." *Administrative Science Quarterly* 31 (3): 439–456.

Tushman, M., and J. Murmann. 1998. "Dominant Designs, Technology Cycles, and Organizational Outcomes." *Research in Organizational Behavior* 20: 231–266.

Tushman, M., and E. Romanelli. 1985. "Organizational Evolution: A Metamorphosis Model of Convergence and Reorientation." *Research in Organizational Behavior* 7: 171–222.

Tushman, M., and L. Rosenkopf. 1992. "On the Organizational Determinants of Technological Change: Towards a Sociology of Technological Evolution." *Research in Organizational Behavior* 14: 311–347.

Ulrich, K. 1995. "The Role of Product Architecture in the Manufacturing Firm." *Research Policy* 24 (3): 419–440.

UN Conference on Trade and Development (UNCTAD). 2006. "Review of Maritime Transport, 2006." Accessed April 23, 2012. http://www.unctad.org/en/docs/rmt2006_en.pdf.

United Nations. 1986. *Transnational Corporations in the International Semiconductor Industry.* Technical Report. New York: United Nations Centre on Transnational Corporations.

U.S. Department of Energy. 2010. "$1/W Photovoltaic Systems: White Paper to Explore a Grand Challenge for Electricity from Solar." http://www1.eere.energy.gov/solar/sunshot/pdfs/dpw_white_paper.pdf.

Utterback, J. 1994. *Mastering the Dynamics of Innovation*. Boston: Harvard Business School Press.

van den Ende, J., and W. Dolfsma. 2005. "Technology-Push, Demand-Pull and the Shaping of Technological Paradigms." *Journal of Evolutionary Economics* 15 (1): 83–99.

Vernon, R. 1971. *Sovereignty at Bay*. New York: Basic Books.

Vincenti, W. 1994. "Variation-Selection in the Innovation of the Retractable Airplane Landing Gear: The Northrop 'Anomaly.'" *Research Policy* 23 (5): 575–582.

von Burg, U. 2001. *The Triumph of Ethernet: Technological Communities and the Battle for the LAN Standard*. Stanford, CA: Stanford University Press.

von Hippel, E. 1988. *The Sources of Innovation*. New York: Oxford University Press.

von Tunzelmann, N. 1978. *Steam Power and British Industrialization to 1860*. Oxford: Oxford University Press.

Waits, R. 2000. "Evolution of Integrated-Circuit Vacuum Processes: 1959–1975." *Journal of Vacuum Science Technology* 18 (4): 1736–1745.

Walker, R. 1992. *Silicon Destiny: The Story of Applications Specific Integrated Circuits and LSI Logic Corporation*. Palo Alto, CA: CMC Publications.

Walsh, V. 1984. "Invention and Innovation in the Chemical Industry: Demand-Pull or Discovery-Push?" *Research Policy* 13 (4): 211–234.

Watkinson, J. 1999. "The History of Digital Audio." In *Magnetic Recording: The First 100 Years*, edited by E. Daniel, C. Mee, and M. Clark, 110–123. New York: IEEE Press.

Wikipedia. 2009a. "Dynamic Random Access Memory." Accessed July 17, 2009. http://en.wikipedia.org/wiki/DRAM.

———. 2009b. "Field Programmable Gate Array." Accessed July 17, 2009. http://en.wikipedia.org/wiki/Field_programmable_gate_array.

———. 2009c. "Transistor Count." Accessed July 18, 2012. http://en.wikipedia.org/wiki/Transistor_count.

Williamson, R. 2010. "Strategies for the Future of Lighting." Unpublished MS thesis, Massachusetts Institute of Technology, Cambridge, MA.

Windrum, P. 2005. "Heterogeneous Preferences and New Innovation Cycles in Mature Industries: The Amateur Camera Industry 1955–1974." *Industrial and Corporate Change* 14 (6): 1043–1074.

Winter, S. 2008. "Scaling Heuristics Shape Technology! Should Economic Theory Take Notice?" *Industrial and Corporate Change* 17 (3): 513–531.

Yang, H., C. Phelps, and H. Steensma. 2010. "Learning from What Others Have Learned from You: The Effects of Knowledge Spillovers on Originating Firms." *Academy of Management Journal* 53 (2): 371–389.

Yinug, F. 2007. "The Rise of the Flash Memory Market: Its Impact on Firm Behavior and Global Semiconductor Trade Patterns." *Journal of International Commerce and Economics* (July). http://www.usitc.gov/publications/332/journals/rise_flash_memory_market.pdf.

Yoffie, D. 1993. "The Global Semiconductor Industry, 1987." Harvard Business School Case 9-388-052.

Yoffie, D., P. Yin, and E. Kind. 2004. "QUALCOMM, Inc. 2004." Harvard Business School Case 705-401.

Yu, D., and C. Hang. 2011. "Creating Technology Candidates for Disruptive Innovation: Generally Applicable R&D strategies." *Technovation* 31 (8): 401–410.

Zervos, A. 2008. "Status and Perspectives of Wind Energy." In *IPCC Scoping Meeting on Renewable Energy Sources—Proceedings*, edited by O. Hohmeyer and T. Trittin, 103–125. Geneva: Intergovernmental Panel on Climate Change.

Index